Modern Methods
of
Pipe Fabrication

By

S.D. BOWMAN

AUTHOR OF

ORDINATES FOR 1000 PIPE INTERSECTIONS
SELECTED PIPING PROBLEMS

THIRTEENTH PRINTING

CLAITOR'S PUBLISHING DIVISION
3653 Perkins Rd., P.O. Box 261333
Baton Rouge, LA 70826-1333

2015

The Author: S.D. Bowman
PO BOX 86651
Baton Rouge, LA 70879
Tel: (225) 937-2965
 email: stepbow1@att.net

The publisher: CLAITOR'S PUBLISHING DIVISION
3653 Perkins Rd., P.O. Box 261333
Baton Rouge, LA 70826-1333
Tel: 800-274-1403 (In LA 225-344-0476)
Fax: 225-344-0480
email: claitors@claitors.com
Order Form/URL: www.claitors.com

MODERN METHODS OF PIPE FABRICATION

The author, S. D. Bowman, has had twenty years of experience in industrial pipe work. This includes pipefitting, fabrication, layout, and several years as an instructor of pipe apprentices.

He has compiled this **manual** specifically for use by the pipe fabricator in figuring out piping offsets and intersections in the easiest, most accurate, and economical manner. It contains information most likely to be needed in his everyday work and omits unnecessary procedures and explanations.

The application of the right triangle to *all* piping offsets and intersections is a tried and tested practical method of solving piping problems. It is rapidly becoming a standard procedure with pipefitters everywhere.

TABLE OF CONTENTS

SOLUTION OF RIGHT TRIANGLES

The sum of the three angles of any triangle is 180°. A right triangle has one 90° angle and two acute angles (angles less than 90°) whose sum is 90°.

For example, if one of the acute angles of a right triangle is 45°, then the other one is 45°. If one of the acute angles of a right triangle is 30°, then the other one is 60°. See figure 1.

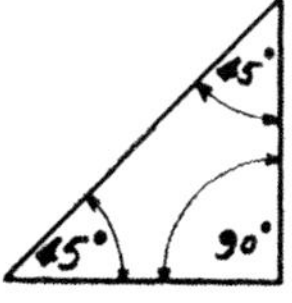

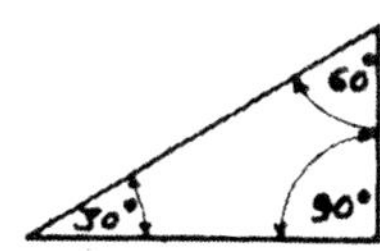

Figure 1
Right Triangles

To solve a right triangle means to find the length of its unknown sides and the degree of its unknown angles. In order to solve any right triangle, there must be certain knowns:

   (1) The length of two sides (fig. 2)

          or

   (2) The degree of two angles (including the 90° angle) and the length of one side (fig. 3)

If these minimums are not known, then the triangle cannot be solved. However, in pipe work this information is given or can be found.

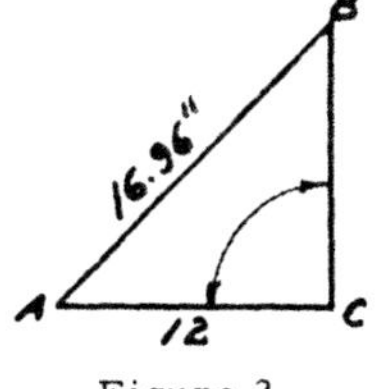

Figure 2

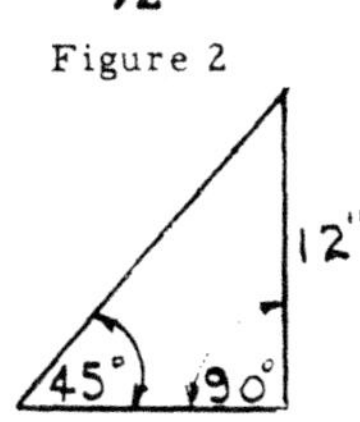

Figure 3

Figure 4 shows the names of the principal center to center measurements involved in a double offset. The offset is merely shortened to "Set." The "Travel" is the distance that the angled piping takes up or travels straight ahead in the pipe line. And

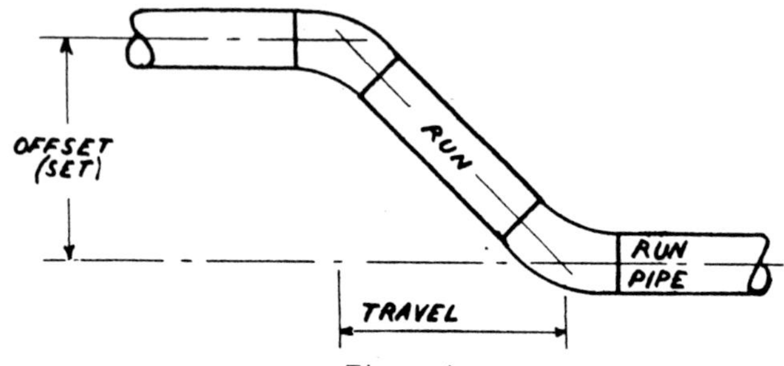

Figure 4

the "Run" is the pipe that connects the parallel ends of the pipe to offset the line.

Figure 5 illustrates how a right triangle is formed by connecting the limits of the set, run, and travel. This is one of the many applications of the right triangle to piping problems.' It can be seen that to completely dimension the different parts of the offset and subsequently obtain the cut lengths of each piece of pipe, the length of the run and travel must be found. This is done by solving the right triangle thus formed.

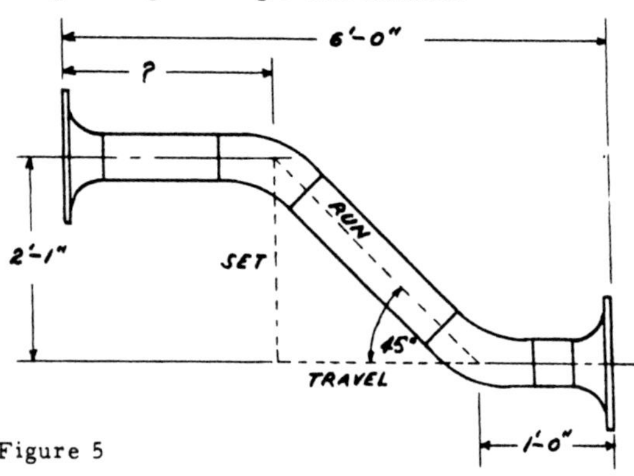

Figure 5

Right triangles are solved by finding the ratio of the different sides to each other. These sides are named and correspond to the run, set, and travel of an offset. This is illustrated in figure 6.

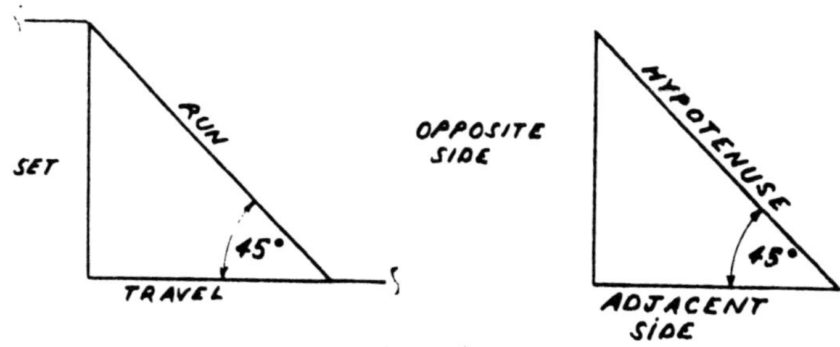

Figure 6

It can be seen that the run and hypotenuse, the set and the opposite side, and the travel and the adjacent side correspond to each other in the two triangles.

How to Identify the Sides
of a Triangle by Name

The hypotenuse of a right triangle is easy to recognize because it is always across from the 90° angle. The opposite side is opposite from the angle you are referring to such as the 30° angle in figure 7. The adjacent

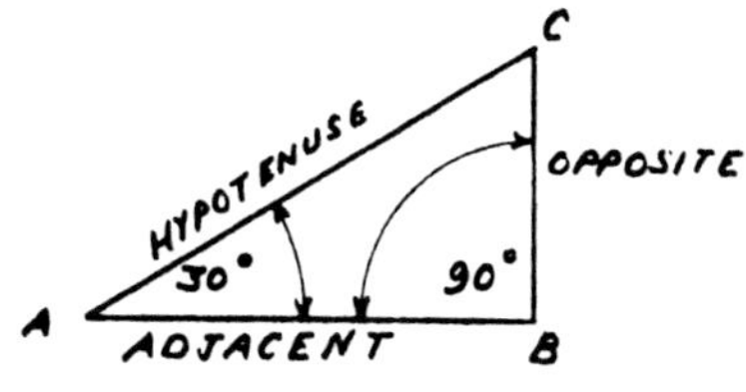

Figure 7
Sides of a right triangle

side is next to or adjacent to the acute angle of 30° in triangle ABC.

It can be said, then, that two of the sides of a right triangle (opposite and adjacent) are named in relation to their acute angles. This is illustrated in figure 8. Notice that the hypotenuse is

3

always across from the 90° angle.

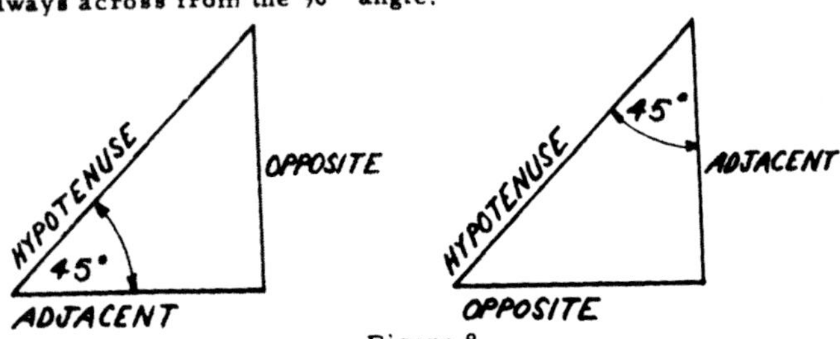

Figure 8

Names of sides as related to acute angles

How to Find the Degree of Angle when Length of Two Sides is Given

The ratio of the sides of the triangle to each other is found by using the following formulas:

FORMULAS FOR FINDING FUNCTIONS OF ANGLES

$$\frac{\text{Side Opposite}}{\text{Side Adjacent}} = \text{Tangent} \qquad \frac{\text{Hypotenuse}}{\text{Side Adjacent}} = \text{Secant}$$

$$\frac{\text{Side Adjacent}}{\text{Side Opposite}} = \text{Cotangent} \qquad \frac{\text{Hypotenuse}}{\text{Side Opposite}} = \text{Cosecant}$$

The known lengths of the sides of the triangle are substituted in the appropriate formula and the indicated division is performed. The result is looked up in the table of natural trignometric functions. The degree that corresponds to that function is the angle included between those two sides of the triangle.

Example problem: Find angle of fitting when run and offset are given.

It is evident that the hypotenuse is 40" and the side opposite the angle of the fitting is 24" (figure 9). Select the formula that has these two knowns. It is:

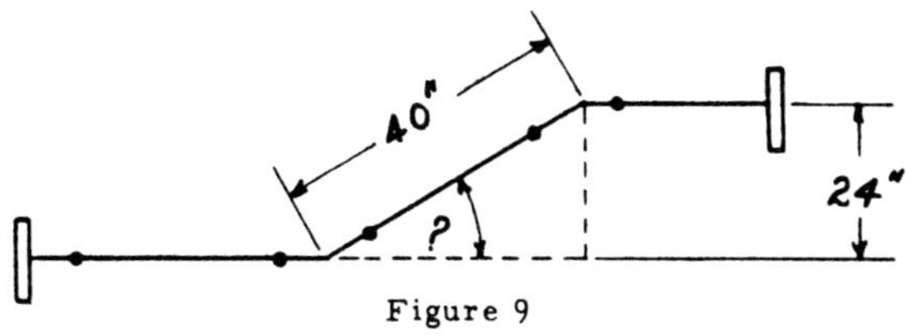

Figure 9

$$\frac{\text{Hypotenuse}}{\text{Opposite}} \quad = \quad \text{Cosecant of the angle}$$

Substituting the known sides in the formula:

$$\frac{40}{24} \quad = \quad 1.666 \quad = \quad \text{Cosecant of the angle}$$

From the table (figure 10) the degree that has 1.666 for its cosecant is 36° - 53'. This is the angle of the fitting necessary for a 24" offset when the run pipe measures 40" from center to center. This degree is found by looking through the table under the column headed "cosecant" until you come to the number 1.666. If this exact number cannot be found, then use the number closest to it.

How to read the table

Each degree in the table is divided into 60 parts called minutes. When reading the table, the degrees from 0 through 44 are listed at the top of the page and their minutes are read downward in the column on the extreme left side. Degrees 45 through 89 are listed on the bottom of the page and their minutes are read upward in the column on the extreme right hand side. Examine the page from the trigonometry tables (figure 10). As an example, the sine of 53°-15' is .80125; the tangent of 36°-30' is .73996.

In three of the columns there is a whole number followed by a four place decimal fraction. Notice that the whole number is indicated only in every fifth number of those columns. When

36°

M	Sine	Cosine	Tan.	Cotan.	Secant	Cosec.	M
0	.58778	.80902	.72654	1.3764	1.2361	1.7013	60
1	.58802	.80885	.72699	.3755	.2363	.7006	59
2	.58825	.80867	.72743	.3747	.2366	.6999	58
3	.58849	.80850	.72788	.3738	.2368	.6993	57
4	.58873	.80833	.72832	1.3730	.2371	.6986	56
5	.58896	.80816	.72877	1.3722	1.2374	1.6979	55
6	.58920	.80799	.72921	.3713	.2376	.6972	54
7	.58943	.80782	.72966	.3705	.2379	.6965	53
8	.58967	.80765	.73010	.3697	.2382	.6959	52
9	.58990	.80747	.73055	.3688	.2384	.6952	51
10	.59014	.80730	.73100	1.3680	1.2387	1.6945	50
11	.59037	.80713	.73144	.3672	.2389	.6938	49
12	.59060	.80696	.73189	.3663	.2392	.6932	48
13	.59084	.80679	.73234	.3655	.2395	.6925	47
14	.59107	.80662	.73278	.3647	.2397	.6918	46
15	.59131	.80644	.73323	1.3638	1.2400	1.6912	45
16	.59154	.80627	.73368	.3630	.2403	.6905	44
17	.59178	.80610	.73412	.3622	.2405	.6898	43
18	.59201	.80593	.73457	.3613	.2408	.6891	42
19	.59225	.80576	.73502	.3605	.2411	.6885	41
20	.59248	.80558	.73547	1.3597	1.2413	1.6878	40
21	.59272	.80541	.73592	.3588	.2416	.6871	39
22	.59295	.80524	.73637	.3580	.2419	.6865	38
23	.59318	.80507	.73681	.3572	.2421	.6858	37
24	.59342	.80489	.73726	.3564	.2424	.6851	36
25	.59365	.80472	.73771	1.3555	1.2427	1.6845	35
26	.59389	.80455	.73816	.3547	.2429	1.6838	34
27	.59412	.80437	.73861	.3539	.2432	.6831	33
28	.59435	.80420	.73906	.3531	.2435	.6825	32
29	.59459	.80403	.73951	.3522	.2437	.6818	31
30	.59482	.80386	.73996	1.3514	1.2440	1.6812	30
31	.59506	.80368	.74041	.3506	.2443	.6805	29
32	.59529	.80351	.74086	.3498	.2445	.6798	28
33	.59552	.80334	.74131	.3489	.2448	.6792	27
34	.59576	.80316	.74176	.3481	.2451	.6785	26
35	.59599	.80299	.74221	1.3473	1.2453	1.6779	25
36	.59622	.80282	.74266	.3465	.2456	.6772	24
37	.59646	.80264	.74312	.3457	.2459	.6766	23
38	.59669	.80247	.74357	.3449	.2461	.6759	22
39	.59692	.80230	.74402	.3440	.2464	.6752	21
40	.59716	.80212	.74447	1.3432	1.2467	1.6746	20
41	.59739	.80195	.74492	.3424	.2470	.6739	19
42	.59762	.80177	.74538	.3416	.2472	.6733	18
43	.59786	.80160	.74583	.3408	.2475	.6726	17
44	.59809	.80143	.74628	.3400	.2478	.6720	16
45	.59832	.80125	.74673	1.3392	1.2480	1.6713	15
46	.59856	.80108	.74719	1.3383	.2483	.6707	14
47	.59879	.80090	.74764	.3375	.2486	.6700	13
48	.59902	.80073	.74809	.3367	.2488	.6694	12
49	.59926	.80056	.74855	.3359	.2491	.6687	11
50	.59949	.80038	.74900	1.3351	1.2494	1.6681	10
51	.59972	.80021	.74946	.3343	.2497	.6674	9
52	.59995	.80003	.74991	.3335	.2499	.6668	8
53	.60019	.79986	.75037	.3327	.2502	.6661	7
54	.60042	.79968	.75082	.3319	.2505	.6655	6
55	.60065	.79951	.75128	1.3311	1.2508	1.6648	5
56	.60088	.79933	.75173	.3303	.2510	.6642	4
57	.60112	.79916	.75219	.3294	.2513	.6636	3
58	.60135	.79898	.75264	.3286	.2516	.6629	2
59	.60158	.79881	.75310	.3278	.2519	.6623	1
60	.60181	.79863	.75355	1.3270	1.2521	1.6616	0

M	Cosine	Sine	Cotan.	Tan.	Cosec.	Secant	M

53°

reading numbers in these columns, always include the whole number before the decimal, even though it is not shown. They were omitted for the purpose of making the table easier to read.

<u>How to find the length of Run, Set and Travel</u>

The length of the run, set and travel of an offset are also found by solving the right triangle. This is done by selecting one of the formulas below and substituting in it the known side and function of the angle of turn of the offset. The designated multiplication is then performed to obtain the length of the unknown side.

FORMULAS FOR FINDING THE LENGTH OF SIDES FOR RIGHT TRIANGLES WHEN ONE ACUTE ANGLE AND ONE SIDE IS KNOWN

Length of side opposite = Hypotenuse X sine
Side adjacent X tangent   or

Length of side adjacent = Hypotenuse X cosine
Side opposite X cotangent   or

Length of hypotenuse = Side opposite X cosecant
Side adjacent X Secant   or

Example problem: Find length of run pipe when amount of offset and degree of turn is given.

The angle is 50° and the side opposite is 18", so pick out the formula that has those knowns in it. It is:

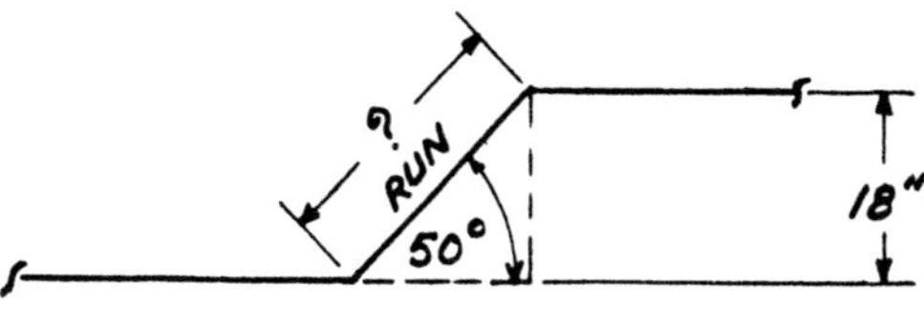

Figure 11

Length of hypotenuse (run) = side opposite X cosecant of the angle
Substituting in the formula:
Length of hypotenuse = 18" X 1.305 (cosecant of 50°)
Length of hypotenuse = 23.49" which is length of run pipe

Example problem: Find the length of travel when amount of offset and degree of turn is given.

The angle is 50°
and the side opposite
is 18" so select the
formula that has
those knowns in it.
It is:

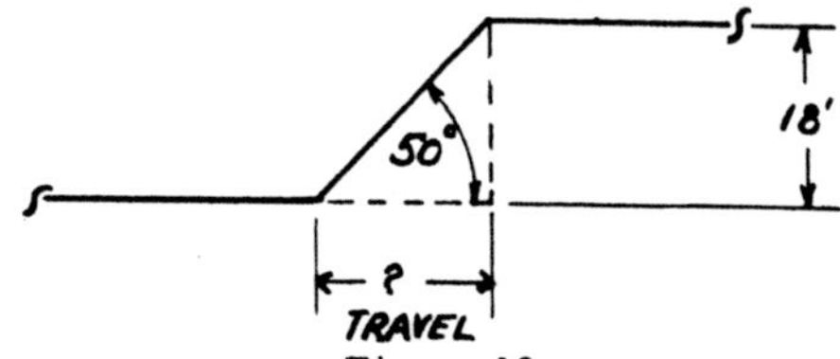

Figure 12

Length of adjacent side (travel) = side opposite X cotangent
of the angle.

Substituting in the formula:

Length of adjacent side = 18" X .8391 (cotangent of 50°)
Length of adjacent side = 15.10" which is length of travel of
the offset.

Example problem: Find the amount of offset when the length
of the run and the degree of turn is given.

The hypotenuse
of the triangle is
23.5" and the de-
gree of turn of the
fitting is 50°. Se-
lect the formula
that has those knowns
in it. It is:

Figure 13

Length of opposite side (offset) = hypotenuse X sine of the angle

Substituting in the formula:

Length of opposite side = 23.5" X .766 (sine of 50°)
Length of opposite side = 18" which is the amount of offset

8

<u>How to find the take up of welding elbows other than 90° turns.</u>
(Take up means center to face dimension)

The take up of 90° standard radius elbows is 1 1/2 times their nominal pipe size. For example a 6" 90° elbow takes up 9" in the pipe line. The take up of short radius elbows is the same as their nominal pipe size. Therefore a 6" short radius welding elbow takes up 6" in the line. But the take up of elbows other than 90° is found by solving a right triangle.

The elbow is sketched and bisected by line AB to find the center point "B". Angle A of triangle ABC is 1/2 of the degree of the elbow. the object is to find the <u>opposite</u> <u>side</u> BC of triangle ABC, which is the take up of the fitting.

Example problem: Find take up of a 50° welding elbow, standard radius, size 8".

Sketch the elbow, bisect the fitting with line AB, and form the triangle ABC as shown in figure 15.

Side BC is the take up of the fitting and the opposite side of angle A in triangle ABC. Side AC is the radius of the fitting (12") and angle A is 25°......one-half of the degree of turn.

From the formulas for finding sides of triangles when one side and one acute angle is given, select one having the above mentioned knowns. It is:

Length of side opposite = side adjacent X tangent

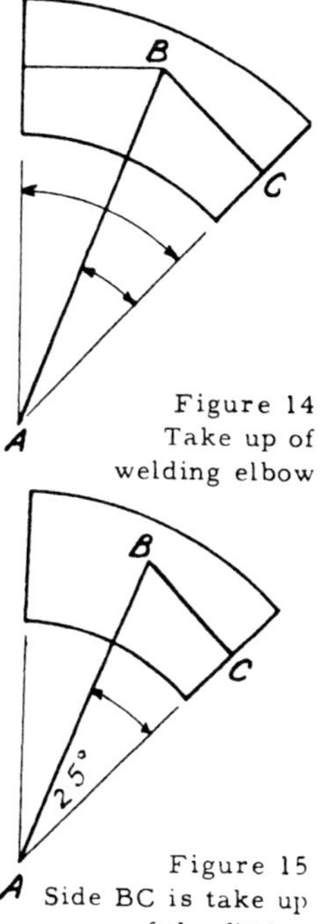

Figure 14
Take up of
welding elbow

Figure 15
Side BC is take up
of the fitting

Length of side opposite $\doteq$ 12" X .4663
Length of side opposite = 5.59" which is the take up of the 50°
elbow

The take up of all welding elbows other than 90° turns are figured in exactly the same way. This procedure can be stated as a formula: <u>Take up of elbow equals tangent of one-half the degree of turn times radius of the elbow.</u>

<u>How to find cut lengths of pipe for dimensioned offset</u>

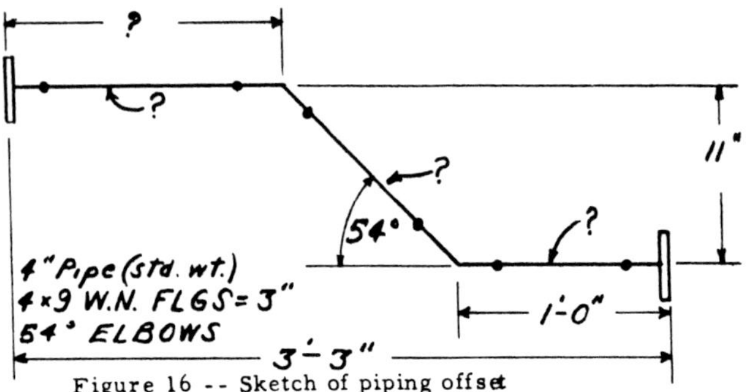

Figure 16 -- Sketch of piping offset

Figure 16 shows a typical piping offset with the dimensions and specifications that are usually given. The object is to find the cut lengths of the three pieces of pipe necessary to fabricate the offset so that it will meet these specifications.

The first step is to find the unknown center to face of flange dimension so the cut length to make that end of the offset can be found. This is done by finding the travel of the offset (adjacent side of the triangle (figure 17), adding it to the 1'-0" dimension given, and subtracting their sum from the overall dimension 3'-3".

Use the formula for finding the adjacent side when one side and one acute angle of the triangle is known:

10

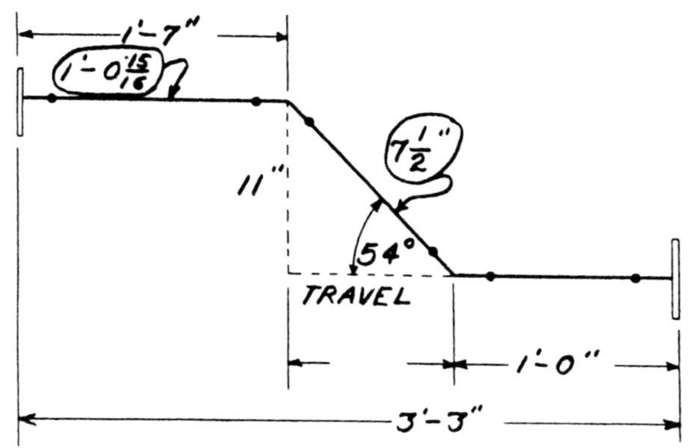

Figure 17

Length of side adjacent = Side opposite times cotangent

Substituting in the formula:

Side adjacent = 11" times .7265
Side adjacent = 7.99" or 8" which is the travel of the offset

Travel of the offset (8") + Center to end dimension (1'-0") =
1'-8"

3'-3" Face to face dimension of the offset
- 1'-8"
1'-7" Center to end dimension of the offset

Next, find the cut length of the pipe for the 1'7" end of the offset. To do this it is necessary to find the take up of the 54° welding elbow. The take up is then added to the flange center to face dimension and their sum subtracted from the 1'-7" end of the offset to obtain that cut length.

Take up of elbow = Tangent of one-half the degree of turn times the radius of the elbow.

Take up of elbow = Tangent of 27° times radius of elbow

Take up of elbow = .5095 X 6" = 3.05" or 3 1/16"

    3 1/16" Take up of elbow
  +3   " Take up of flange
  ‾‾‾‾‾‾‾‾
    6 1/16"

    1'-7"
  -    6 1/16"
  ‾‾‾‾‾‾‾‾
    1'-0 15/16" Cut length of pipe for 1'-7" end of offset.

Next, find the cut length of the piece of pipe that goes between the two 54° elbows. This is done by finding the center to center length of the run pipe and deducting the take up of the two elbows from it.

Using the formula:

Hypotenuse (run) = Opposite side (offset) X Cosecant

Hypotenuse = 11" X 1.236

Hypotenuse = 13.6" or 13 5/8" which is the length of the run
pipe

    13 5/8"
  - 6 1/8" (take up of two 54° elbows previously determined)
  ‾‾‾‾‾‾‾‾
    7 1/2" cut length of pipe to go between the two elbows.

The cut length of the pipe to go between the flange and the 54° elbow on the 1'-0" end of the offset is found by subtracting the take up of both fittings from that dimension. This is the same as was done for the other end of the offset.

    3   " (flange take up)   1'-0"
  +3 1/16" (54° elbow take up)   -   6 1/16"
  ‾‾‾‾‾‾‾‾   ‾‾‾‾‾‾‾‾
    6 1/16"   5 15/16" cut length of pipe
                                                for 1'-0" end of
                                                offset

Notice that the cut lengths are written near their corresponding segment of the offset and are encircled. This makes them easier to see and less likely to be confused with other dimensions on the drawing.

1. How to lay off two 45° elbows from one 90° elbow

Measure the back center of the 90° elbow from beveled edge to beveled edge. Divide this distance in half and mark the elbow at that point. For measuring, use a steel tape held flat against the fitting and make the mark with a sharpened soapstone crayon.

Measure and mark the center of the throat in the same manner.

Set degree protractor, with blade extended, on 45° angle.

With the elbow lying flat, place a framing square on it with a blade touching both sides of each end. See figure 18. Hold the square near the top side of the elbow.

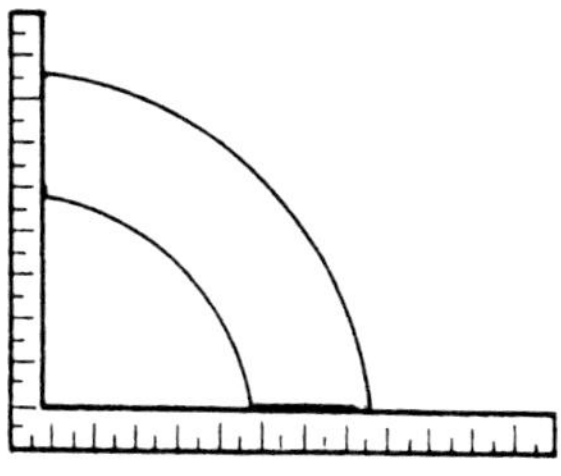

Figure 18

With framing square in position, slide straight edge of protractor frame along outside edge of framing square until blade crosses the vertex of the angle formed by the inside edge of the tongue and blade of the square. See figure 19. Mark the top side of the elbow where the blade of the protractor crosses it. Then turn the elbow over and mark the other side in the same manner.

Align a flexible straight edge (such as a steel tape measure or

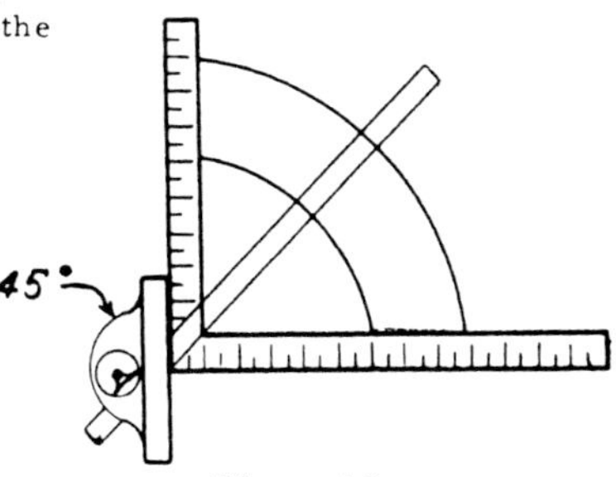

Figure 19

clock spring) with the throat and back of elbow marks and with the mark on top of the elbow. Connect these points into a cutting line using a sharpened soapstone crayon. Turn the elbow over and mark the other side in the same manner.

This cutting line divides the elbow into two 45° parts for a straight-in cut. If it is desired to have the 45° elbows cut "on the bevel," then step back the length of the bevel on each side of this line. Scribe those lines just as the original line was marked.

2. How to lay off three 30° elbows from one 90° elbow.

The procedure is the same as for 45° elbows except that the back and the throat of the elbow is divided into 3 equal parts and the protractor is set on 30°.

3. How to lay off 90° elbow to make turns other than 45 and 30 degrees.

There are three ways to lay off a 90° elbow for cutting into turns of lesser degrees, any one of which can be used. However, the ratio and proportion method is generally considered to be the easiest as well as fastest to use.

Example problem: Lay off a 6" 90° welding elbow to obtain a 54° elbow. Use ratio and proportion method.

Measure length of back of elbow from beveled edge to beveled edge. (It measures 19 1/4".) This length is for 90° of turn, therefore a proportion can be set up:

Length of back of elbow is to 90° as unknown length is to 54°. This is written down as:

$$19.25" \; : \; 90° \; = \; x \; : \; 54°$$
$$90x \; = \; 1039.50$$
$$x \; = \; 11.55" \text{ or } 11 \; 9/16" \text{ Length of}$$

back of elbow required for 54° turn

The length to mark the throat is determined in the same manner.

Measure these lengths from the beveled edge of the elbow,

using a steel tape held flat against the elbow. Mark with sharpened soapstone crayon.

Set degree protractor, with blade extended, on 54°.

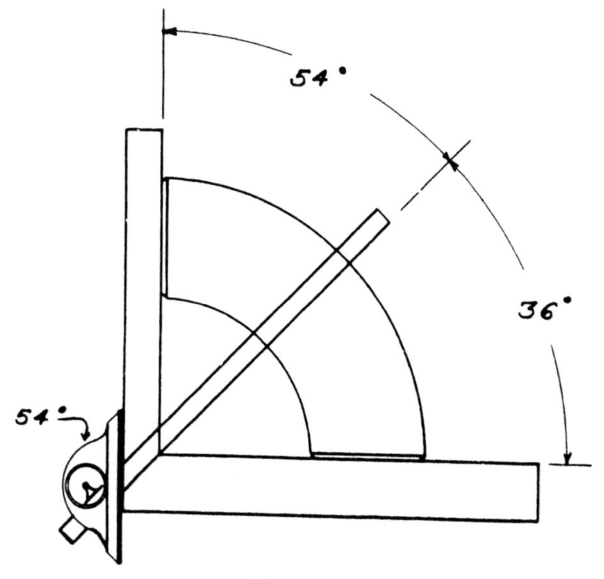

Figure 20

Apply the framing square and degree protractor to the elbow
as shown in figure 20. Hold the framing square approximately
1" from the top of the elbow so that the protractor blade will
touch the top of the elbow as it crosses it. When marking large
elbows, a long protractor blade must be used. If one is not
available, a short straight edge can be used to extend the protractor blade so it will reach the center of the elbow. Mark the
top of the elbow with soapstone crayon, using the protractor
blade as a guide. Then turn it over and mark the other side

in the same manner, making sure that both 54° marks are on the same end of the elbow.

Connecting the marks into a cutting line

Align a flexible straight edge with the throat and back of elbow marks and with the mark on the top side. Connect these points into a cutting line using soapstone crayon to mark with. Turn the elbow over and mark the other side in the same manner. The cutting line divides the 90° elbow into 54° and 36° parts.

> Example problem #2: Lay off a 6" 90° elbow to obtain a 54° and a 36° elbow. Use formula for finding the arc of a circle.

A 90° elbow is one-fourth of a circle. Therefore the formula for finding the arc of a circle can be applied to determine the length that will be contained within a given part of the elbow. Actually, two arcs are found . . . . one in the throat and one on the back center.

The formula for each of these arcs is stated as follows:

Length of arc on back of elbow = Sine of 1° X number of degrees X outside radius

Length of arc in throat of elbow = Sine of 1° X number of degrees X inside radius

Note: Outside radius is the radius of the elbow measured to the back of the elbow. It is radius of the elbow plus one-half of the outside diameter of the elbow. (Figure 21)

Inside radius is the radius of the elbow measured to the throat of the elbow. It is radius minus one-half of the outside diameter of the elbow.

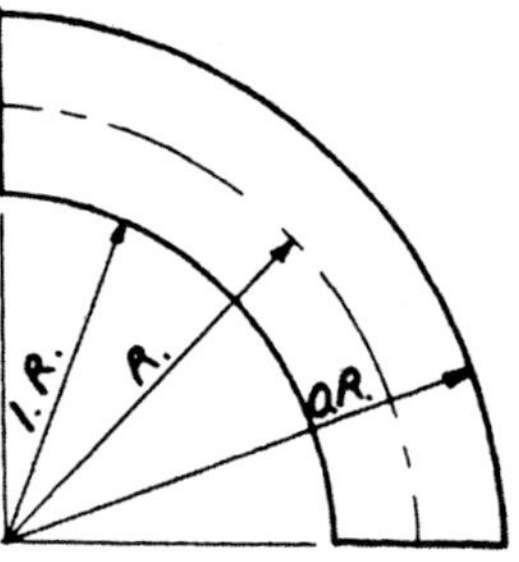

Figure 21

Substituting in the formula:

Arc length on back of elbow = .01745 X 54 X 12.312 (9" + 3.312")
Arc length on back of elbow = 11.59" or 11 5/8"

Substituting in the other formula:

Arc length in throat of elbow = .01745 X 54 X 5.688" (9 - 3.312")
Arc length in throat of elbow = 5.35" or 5 3/8"

Mark the sides of the elbow by using a framing square and degree protractor as shown in figure 20. Connect the four points into a cutting line as described in example problem # 1 to complete the layout.

> Example Problem # 3: Lay off 6" 90° elbow to obtain a 54° elbow. Find length of arc for 1° and multiply it by the number of degrees wanted in the turn.

Measure length of back of 90° elbow from beveled edge to beveled edge. This length (19 1/4") is for 90° of turn. To obtain length of arc for 1°, divide 19 1/4" by 90°. Multiply the result by number of degrees wanted to obtain length of arc for back of elbow.

```
        .2138" for 1° of elbow          2138
 90 )19.2500                               54
                                         8552
                                        10690
                                        11.5452" for back of 54°
                                                         elbow
```

Determine the length of the arc for the throat of the elbow in the same manner.

Measure these lengths, on the back and in the throat, from the beveled edge of one end of the elbow. Mark with soapstone crayon.

Mark the sides of the elbow by using the framing square and degree protractor as shown in figure 20. Connect the four points into a cutting line as described in Example Problem # 1.

<u>Solution of Rolling Offsets</u>

A rolling offset is a double offset, one end of which is rotated or rolled so as to effect a change in elevation and lateral position of the pipe at the same time. The fittings normally used to make these offsets are 90° or 45° welding elbows. However, any de-ee fitting can be used depending upon the design of the partic-ar installation.

How to find length of run pipe of rolling offset using 90° elbows.

Example problem # 1

12'-0" offset
18'-0" roll
6" pipe

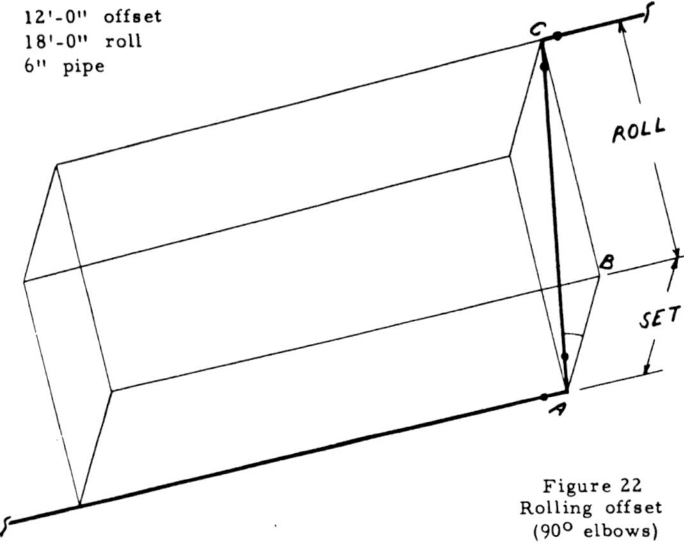

Figure 22
Rolling offset
(90° elbows)

The object is to find side AC of triangle ABC so that the take up of the elbows can be deducted leaving the cut length of the run pipe.

18

It is necessary to find Angle A of triangle ABC before side
AC can be found. Side AB is the adjacent side of angle A in
triangle ABC and BC is the opposite side. Use the formula
for solving right triangles when two sides and one angle are
known. It is:

$$\frac{\text{side opposite}}{\text{side adjacent}} = \text{tangent of the angle}$$

Substituting in the formula: $\frac{18}{12} = 1.500 =$ tangent of angle A

The angle that has 1.500 for its tangent is 56°-19' (found in the
trig tables).

Now that two angles are known (angle A and the right angle)
the run pipe, side AC can be found.

Use the formula for solving a right triangle when one side and
two angles are known. It is:

Length of hypotenuse = opposite side times cosecant

Length of hypotenuse (side AC) = 18 X 1.201

side AC = 21.62' or 21'-7 7/16"

21'-7 7/16"	Length of run pipe
- 1'-6"	Take up of two 6" 90° welding elbows
20'-1 7/16"	Length of cut pipe to go between two elbow

How to find length of run pipe of rolling offset made with 45°
elbows.

Example problem #2

    1'-6" offset
2'-0" roll
6" pipe

The object is to find side AC of triangle ABC, which is the
run pipe of the rolling offset shown in figure 23. When the take
up of the two 45° elbows is deducted from it, the result will be

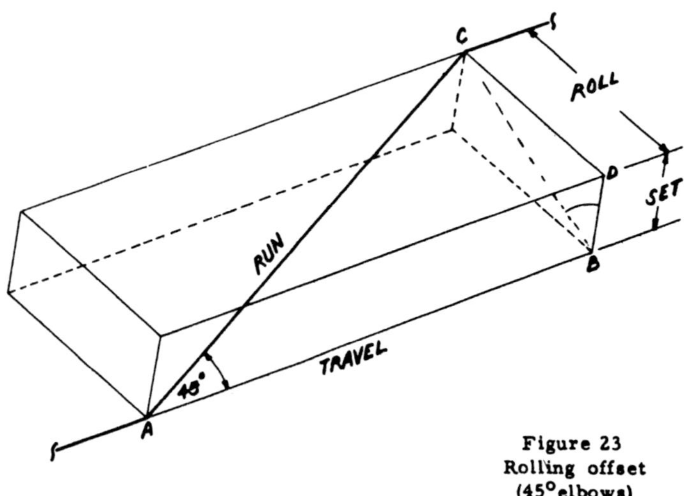

Figure 23
Rolling offset
(45° elbows)

the length of pipe needed to go between the two elbows.

In determining the length of side AC, two triangles must be solved. Side BC of triangle BDC is the <u>true offset</u> of the rolling offset. It's length must be found so that it can be multiplied by the cosecant of angle A to find side AC of triangle ABC.

It is necessary to find angle B of triangle BDC before side BC can be found. Side BD is the adjacent side and DC is the opposite side of angle B. Find angle B by using the formula for solving a right triangle when two sides and the right angle is known. It is:

$$\frac{\text{Side opposite}}{\text{Side adjacent}} = \text{tangent}$$

20

Substituting in the formula:

$$\frac{24}{18} = 1.333 \text{ (Look up under tangent in the trig tables.)}$$

The angle that has 1.333 for its tangent is $53^\circ\text{-}04'$ which is angle B.

Now that angle B is known, side BC of triangle BDC can be found. Use the formula for finding the hypotenuse when the opposite side is known. It is:

Length of hypotenuse = Side opposite X cosecant (of $53^\circ\text{-}04'$)

Substituting in the formula:

Length of hypotenuse = 24 X 1.251
Length of hypotenuse = 30.02"

Side AC can now be found by using the formula for finding the hypotenuse (which is side AC) when the opposite side and two angles are known. It is:

Length of hypotenuse = cosecant X opposite

Substituting in the formula:

Length of hypotenuse = 1.414 X 30.02"
Length of hypotenuse = 42.45"

```
  42.45"
-  7.50   (take up of two 45° elbows)
  34.95   or 2'-10 15/16"  Length of cut of pipe to go between
          the two 45° elbows.
```

The solution to this problem can be stated as a formula: The center to center length of the run pipe of a rolling offset (45° elbows) is determined by finding the length of the hypotenuse of the right triangle formed by the set against the roll and multiplying the result by 1.41.

How to find length of run pipe and degree of turn necessary to make rolling offset where length of travel is given and is such that 45° elbows cannot be used.

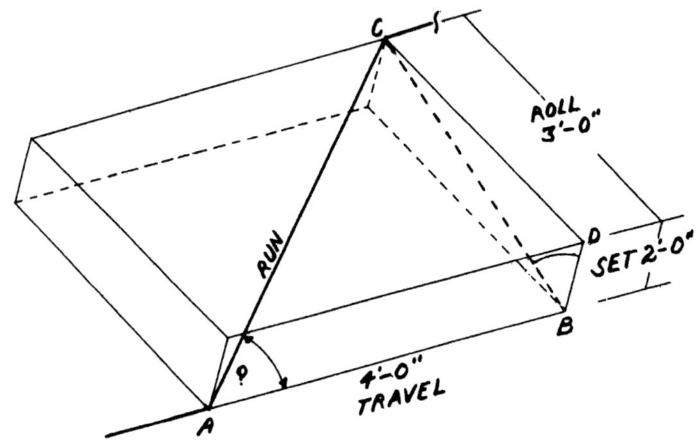

Figure 24
Rolling Offset (travel, set, and roll given)

Example problem #3

   2'0" offset
   3'0" roll
   6" pipe

The only difference between this problem and the preceding one is that after side BC is found, its ratio to side AB is found to determine angle A. The cosecant of Angle A is then multiplied by side BC to find the length of the run pipe AC.

The steps performed to solve this problem are as follows:

      Find angle B
      Find side BC
      Find angle A
      Find side AC which is the length of the run pipe.

It is necessary to find angle B of triangle BDC before side BC can be found. Side BD is the adjacent side and DC is the opposite side when referring to angle B. Find angle B by using the formula for solving a right triangle when two sides and one angle is known. It is:

$$\frac{\text{Side opposite}}{\text{Side adjacent}} = \text{Tangent}$$

Substituting in the formula:

$$\frac{36}{24} = 1.500 \text{ (Look up under tangent in the trig table.)}$$

The angle that has 1.500 for its tangent is 56°-19' and is angle B.

Now that angle B is known side BC of triangle BDC can be found. Use the formula for finding the hypotenuse when the opposite side is known. It is:

Length of hypotenuse = side opposite X cosecant
Length of hypotenuse = 36 X 1.201
Length of hypotenuse = 43.24 or 3-7 1/4" which is side BC.

The degree of angle A must be determined before side AC can be found. Side AB (adjacent side) is 4'-0" and side BC (opposite side) has been found to be 3'-7 1/4". Use the formula for finding an angle when two sides and one angle is known. It is:

$$\frac{\text{Side adjacent}}{\text{Side opposite}} = \text{cotangent}$$

Substituting in the formula:

$$\frac{48}{43.24} = 1.110 \text{ (Look up under cotangent.)}$$

From the trig table, the angle that has 1.110 for its cotangent is 42°-01'. This is angle A.

Now that angle A is known, side AC of triangle ABC can be found. Side BC is the opposite side and has been found to be 43.24". Use the formula for finding the hypotenuse when one side and two angles are known. It is:

Length of hypotenuse = side opposite X cosecant.

Substituting in the formula:

Length of hypotenuse = 43.24 X 1.494
Length of hypotenuse = 64.6" or 5'-4 5/8" center to
center length of run pipe.

The cut length of the run pipe is found by subtracting the take up of the two elbows from the center to center length of the run pipe.

Use the formula for finding the take up of an elbow.

Take up of elbow = tangent of 1/2 the degree of turn X
radius of the elbow.
take up = tangent of 21° X 9" radius
take up = .3838 X 9
= 3.45" or 3 7/16" take up of 42°-01 turn
take up of both elbows =   3 7/16
+3 7/16
―――――
6 7/8"

Length of cut pipe =   5' 4 5/8"
―   6 7/8
―――――――
4' 9 3/4"

MEASURING AND MARKING PIPE FOR BEVELING

When measuring pipe to a given length, emphasis must be placed on reading the rule accurately and marking the pipe correctly. It is always a good idea to double-check your measurements to avoid error and the waste of time and material that often results when pipe is cut too short or too long. If the cut is to be longer than twelve feet, then it is advisable to use a 50-foot steel tape to lessen the chance for error in measuring.

When measuring the cuts of pipe for beveling, the length of the bevel must be subtracted from the overall length of the cut to determine the exact place to mark the cutting line.

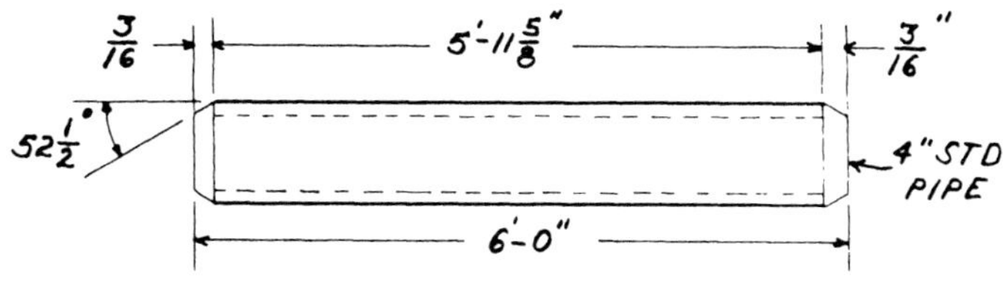

Figure 25

The best method is to mark the exact cut length on the pipe
and "step back" for the bevel take up (figure 25). On standard
weight pipe size 4" and smaller, the bevel length takes up approximately 3/16"; on pipe size 6" through 12", the take up is
1/4". Bevel take up depends upon the thickness of the pipe and
the angle at which the cutting torch is held. The burner should
hold his torch tip at an angle of approximately 52-1/2° from the
run of the pipe to make these beveled cuts.

The cutting line is scribed by using a wrap-around to guide
the crayon around the pipe in a straight line. A wrap-around
is a length of compressed asbestos sheet packing from three
to six inches wide and long enough to encircle the pipe approximately 1-1/2 times. The marking edge of the wrap-around
must be perfectly straight or the device will not be of much
value.

Be sure that the pipe surface is clean where the wrap-around
makes contact with it. This is necessary so that the wrap-
around will take a firm, uniform grip on the pipe - which is
the key to obtaining a straight marking edge. Often it is necessary to burn the protective factory coating off of the pipe in the
place where the wrap-around is to be applied. But, sometimes
it may be more convenient to mark the cutting line and then
center-punch it so the cutting line cannot burn off ahead of the
flame from the cutting torch.

The steps in marking a cutting line are as follows:

1. Measure length of cut and mark with soapstone crayon.

2. Clean surface of pipe several inches on each side of
    cutting mark.

3. Hold one end of wraparound and drape other end over the
    pipe. The straight edge of the wraparound should be even
    with mark on pipe.

4. With one hand, reach and bring end of wraparound up and
    against the pipe. At the same time, lower free end of
    wraparound until the surfaces overlap.

Figure 26

5. Align straight edges and pull tightly on wraparound.
    Mark cutting line with soapstone crayon, using edge
    of wraparound as a guide. (figure 26)

Down through the years, numerous devices have been used as a guide for marking cutting lines on pipe, but none have proved as satisfactory as the wrap-around made from compressed asbestos sheet packing. For better service, this device should be handled carefully on the job and properly stored when not in use. If the marking edges become worn, then the wrap-around should be discarded and a new one obtained.

How to find cut back for mitered turns

On small size pipe it is necessary to use only one cut back length on the top and bottom center line. This cut back length is found by solving for side AB in triangle ABC (figure 27).

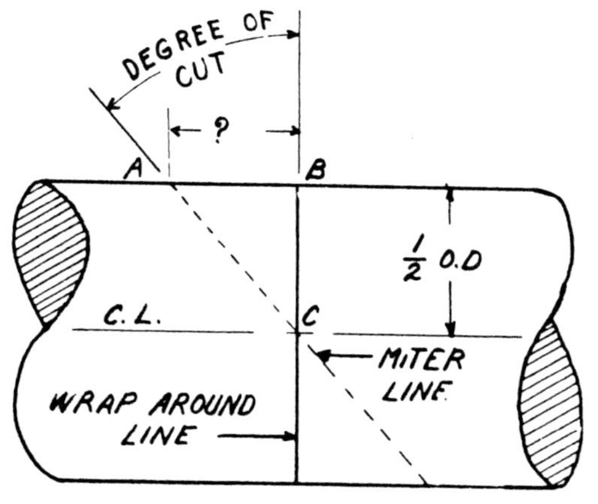

Figure 27
Cut Back for Mitered Turn

Example problem: Find the cut back for 20° turn (3" pipe).

Side AB is the opposite side of angle C in triangle ABC. Angle C is 10°, the angle of the miter cut or 1/2 of the degree of turn.

Side BC is 1.75", which is 1/2 of the outside diameter of the pipe. Side BC is also the adjacent side of the triangle.

Since two angles and one side are known, side AB can be found Use the formula for finding the opposite side when two angles and one side are known.

Length of side opposite = side adjacent X tangent

Substituting in the formula:

Length of side opposite = 1.75" X .1763

Length of side opposite = .308" or 5/16" cut back for 10° miter or 20° angle of turn

This method of finding cut back can be stated as a formula: Cut back for mitered turn = tangent of 1/2 the degree of turn X 1/2 the O.D. of the pipe.

How to Lay off a Miter Cut

1. Mark line around circumference of pipe, using straight edge of wrap-around as a guide. Mark with soapstone crayon. The line should be scribed at the desired center of the turn. (The wrap-around line is used as a reference line from which to measure the cut back lengths. It also locates the side centers of the miter line -- point C in figure 27.)

2. Divide pipe into four equal parts and mark straight lines along run of pipe at these points. These center lines should extend several inches on each side of the wrap-around line.

3. On one of the center lines, measure and mark the cut back length from the wrap-around line.

4. Turn the pipe over (180°) and measure and mark the cut back length on the opposite side of the wrap-around line.

5. Hold a flexible straight edge (such as a steel measuring tape) on one of the cut back points. Let the part of the tape on each side of the pipe cross exactly over each side center point. Mark the cutting line on this half of the pipe, using the flexible straight edge as a guide.

6. Turn the pipe over and repeat the operation to complete the
    layout of the cutting line.

HOW TO FIND CUT BACK ON PIPE WHERE ORDINATE LINES ARE USED.

On the larger size pipe (above 4") it is necessary to divide
the pipe into 8 equal parts and to mark cut back lengths at each
of these places. For pipe above size 10", it is advisable to
further divide the pipe circumference and mark 16 ordinate lines.
This will make for an accurate miter line (cutting line) and
result in a better fit-up.

Figure 28 shows the outside area of the pipe divided into 16
equal parts with ordinate lines drawn to accomodate the cut back
lengths.

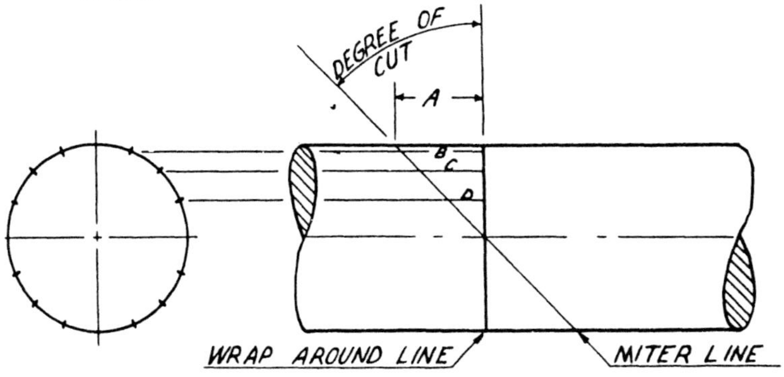

Figure 28
Cut Back Lengths For 16 Ordinate Lines

A formula for finding cut back length for 16 ordinate lines
has been worked out.

Dimension A = Tangent of degree of cut times 1/2 O.D.
                of the pipe. This is the longest cut back.

Dimension B = .923 times A

Dimension C = .707 times A

Dimension D = .382 times A

Dimension A, B, C, and D is repeated in each of four quarters
of the pipe, making 16 ordinate lengths in all. Remember when
using 8 ordinate lines, use dimension A and C only.

For larger size pipe (above 24") it may be desired to use 32
ordinate lines. In this case letter the ordinate lines A thru
H instead of A thru D, and use the following constants to obtain the cut back lengths.

      A = Tangent of degree of cut times 1/2 O.D. of the pipe.
      B = .978 times A
      C = .923 times A
      D = .831 times A
      E = .707 times A
      F = .558 times A
      G = .382 times A
      H = .200 times A

If one-half the outside diameter of the pipe is used when figuring
cut back lengths, then the burner should hold his torch so that he
will make a hacksaw type cut. A radial cut, one that is made by
pointing the torch tip toward the center of the pipe at all times,
is much easier to make. If a radial cut is to be made, then you
must use one-half the inside diameter of the pipe when figuring
cut back lengths.

<u>How to Find Length of Middle Piece of Three Piece 90° Mitered Elbow.</u>

Example problem: Find length of middle piece of 90ᵛ mitered elbow having 12" Radius.

Solution:

Find angle of cut:

Angle of cut =

$$\frac{\text{total number of degrees of turn}}{\text{number of welds times 2}}$$

Angle of cut = $\frac{90}{4}$ = 22 1/2

Therefore angle A of triangle ABC is 22 1/2°.

The radius of the elbow is AB - - - the adjacent side of the triangle.

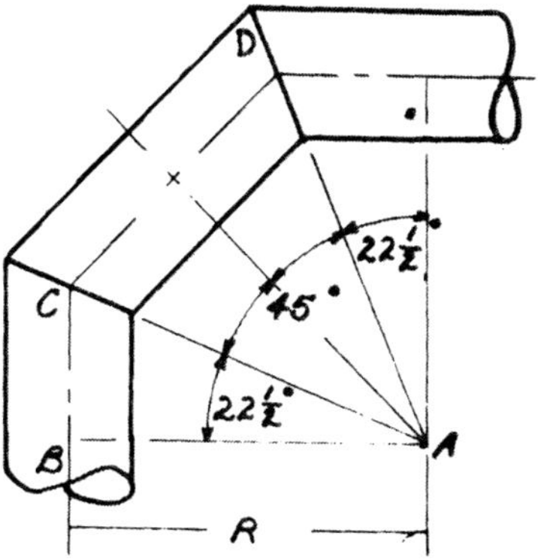

Find side BC by using formula for finding side of triangle when two angles and one side are known:

Length of side opposite = tangent X adjacent
Length of side opposite = .4142 X 12
Length of side opposite = 4.97" which is side BC

Since triangle ACD is composed of two triangles identical to triangle ABC, side CD = 2 X side CB or 9.94".

The solution to this problem can be stated as a formula:

Center to center length of middle piece of 3 piece 90° mitered turn equals 2 X Radius X tangent of angle of cut.

Figure 29
3 Piece 90° Mitered Turn

Given: 4 piece 90° turn,
        6" pipe, 36"
        radius

Object: Find A and B

$$\text{Angle of cut} = \frac{90°}{2 \times \text{No. of welds}}$$

$$\frac{90°}{6} = 15° = \text{Angle of cut}$$

In the small triangles (figure 30) side A is opposite side of a 15° angle. The adjacent side is the 36" radius.

Use the formula:

Length of side opposite =
      tangent X adjacent

A = .26795 X 36   B = 9.63 X 2
A = 9.63"   B = 19.26"

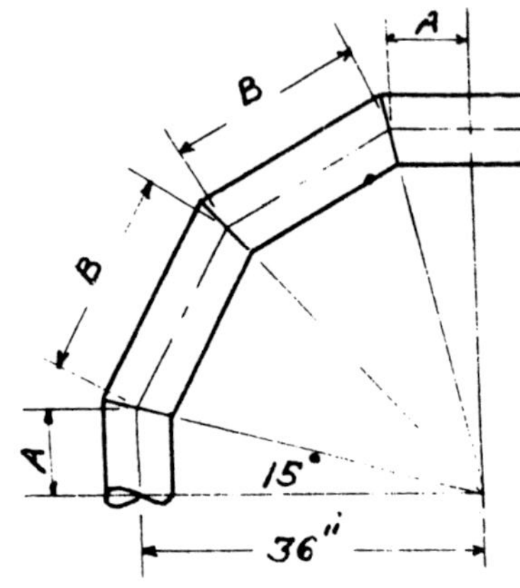

Figure 30

To find the cut back for each miter, use the formula:

Cut back = Tangent of degree of cut X 1/2 the O. D. of
                                        the pipe

Cut back = .26795 X 3.3125
Cut back = .8875 or 7/8" (Cut back on each side of
                                    wrap-around line.)

HOW TO FIND TRAVEL AHEAD NECESSARY TO MAKE EQUAL SPREAD OFFSETS

Each of the center to face lengths of the flanged ends of the three offsets are different lengths (see figure 31). This travel ahead is necessary so that the pipe lines will remain the same distance apart after they are turned to make the offset.

The flanged end of offset # 2 is length BC longer than the same end of offset # 1,

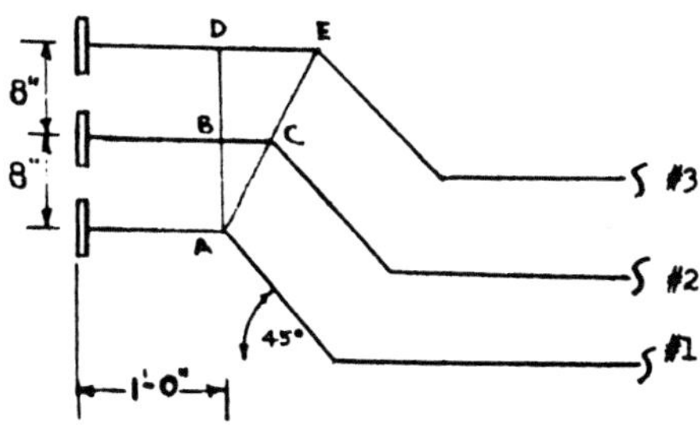

Figure 31

And the end of offset # 3 is length DE longer than the flanged end of offset # 1.

The difference in length is found by solving for side BC in triangle ABC and for side DE in triangle ADE. Each of these dimensions is added to the center to face dimension of offset # 1 to find the center to face length of offsets # 2 and # 3.

Side BC is the opposite side of angle A in triangle ABC. Angle A is one-half the degree of turn or 22 1/2°. Side AB is the adjacent side of angle A.

Length of side opposite (BC) = side adjacent (AB) X tangent

$$BC = 8 \times .4142$$
$$BC = 3.31" \text{ or } 3\ 5/16"$$

3 5/16" plus 1'-0" equals 1' 3 5/16" Center to face length of offset # 2

Side DE is the opposite side of angle A in triangle ADE. Angle A is 22 1/2°.

Side AD is the adjacent side of angle A.

Using the same formula:

    Length of side opposite (DE) = Side adjacent times tangent
                        DE = 16 X .4142
                        DE = 6.62" or 6 5/8"
6 5/8" plus 1'-0" equals 1'-6 5/8" Center to face of offset # 3.

The travel ahead for all equal spread offsets is determined by
using this method.

SPECIAL OFFSETS

In some offsets, the amount of offset will not permit the use of two 45° welding elbows because the run of the offset is less than the take up of the two fittings. And, for reasons of design, an offset is occasionally specified to be made from two welding elbows with no pipe between them...... even though the run pipe would be of sufficient length to use 45° elbows.

To solve problems of this type, the degree of turn of the elbows that will give the desired amount of offset must be found. This is accomplished by solving one triangle, and possibly two, depending upon the type of offset specified. Some examples of these special offsets and their solution are included in this section.

Special Offsets
Example problem # 1

Find degree of elbow required to make offset when amount of offset is more than radius of elbow and it is not desired to use pipe between the fittings.

Object: To find angle A of triangle BAC which will be the degree of each elbow.

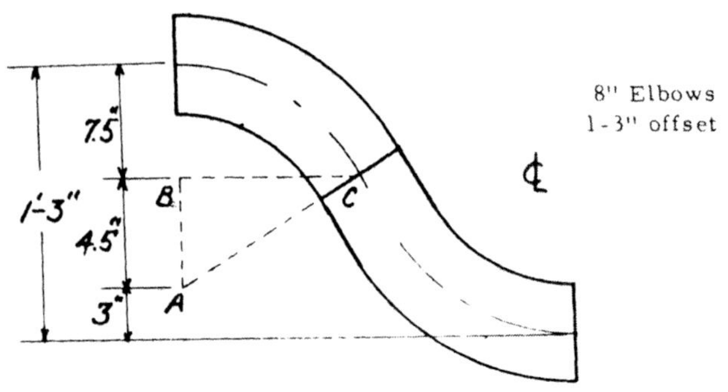

Figure 33

Explanation:

1. Set up problem as shown with center line BC dividing the offset into two 7.5" parts.

2. Side AB of triangle BAC is found by subtracting the difference between amount of offset and radius of elbow from one-half of the offset.

3. Side AC of triangle BAC is the radius of the elbow.

4. Divide side AC by side AB (hypotenuse divided by adjacent side).

5. Look up under <u>secant</u> in table of trigonometric functions. Read the corresponding angle which will be angle A.

In this example problem:

$$\frac{AC}{AB} = \frac{12}{4.5} = 2.666 \quad \text{(Look up under secant.)}$$

From the table of trigonometric functions, the angle that has 2.666 for its secant is 67°-58'. Two elbows, each containing this degree of turn, when buttwelded together will result in 1'-3" of offset.

Special Offsets
Example problem # 2

Find degree of elbows required to make offset when the amount of offset is less than radius of elbow and it is not desired to use pipe between the two fittings.

> Object: To find angle A of triangle BAC which will be the de-
> gree of each elbow

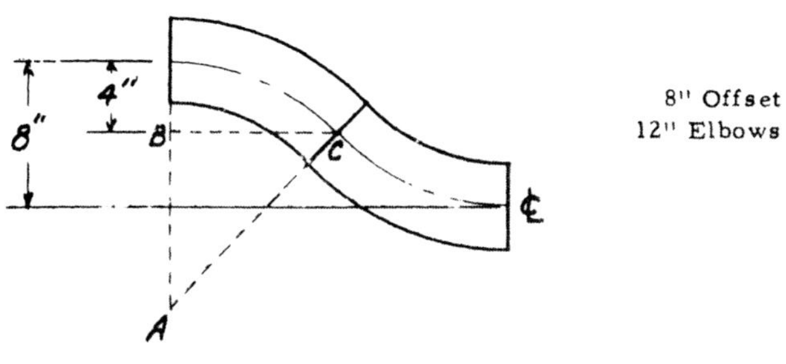

Figure 34

Explanation:

Set up problem with center line dividing the offset into two 4" parts as shown above.

1. Side AB of triangle BAC is found by subtracting one-half of the offset from the radius of the elbow.

37

2. SIDE AC OF TRIANGLE BAC IS THE RADIUS OF THE ELBOW.

3. DIVIDE SIDE AC BY SIDE AB (HYPOTENUSE DIVIDED BY ADJACENT SIDE).

4. LOOK UP RESULT UNDER <u>SECANT</u> IN THE TRIG TABLES. READ THE CORRESPONDING
    ANGLE WHICH WILL BE ANGLE A.

IN THIS EXAMPLE PROBLEM:

$$\frac{AC}{AB} = \frac{18}{14} = 1.285 \ (\text{LOOK UNDER SECANT})$$

FROM THE TRIG TABLES, THE ANGLE THAT HAS 1.285 FOR ITS SECANT IS $38^\circ\text{-}55'$.
TWO ELBOWS, EACH CONTAINING THIS DEGREE, WHEN BUTTWELDED TOGETHER WILL
MAKE AN 8" OFFSET.

SPECIAL OFFSETS
EXAMPLE PROBLEM # 3.

FIND DEGREE TO USE WITH 90° ELBOW TO MAKE A GIVEN OFFSET WHEN THE
AMOUNT OF OFFSET IS LESS THAN THE RADIUS OF THE FITTINGS.

GIVEN:
      6" WELDING ELBOWS (9" RADIUS)

      7½" OFFSET

EXPLANATION:

      SET UP PROBLEM AS SHOWN AT RIGHT
      BY DRAWING RADIUS LINES AND
      CONNECTING LINES TO FORM TRI-
      ANGLES ABC AND ADC.

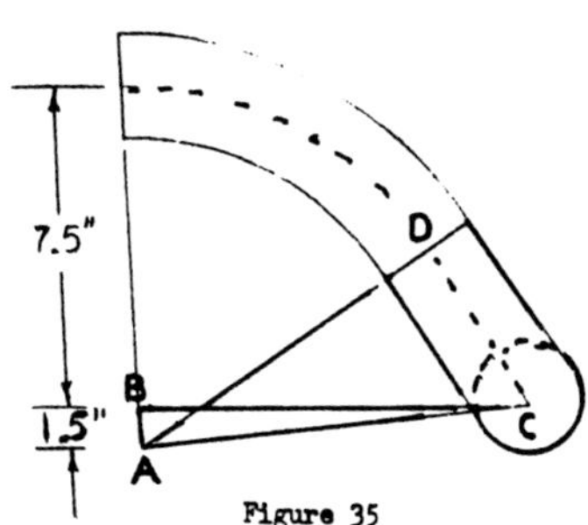

Figure 35

2.   Sides AD and DC of the triangles are the radius of each of the elbows.

3.   Find side AB of triangle ABC by subtracting the amount of offset from the radius of the elbow.

4.   Find side AC of triangle ADC by multiplying the cosecant of 45° times 9.   (Angle A and angle C of triangle ADC are each 45° because the legs of the triangle are of equal length.)

5.   Divide side AC by side AB (hypotenuse divided by adjacent side).

6.   Look up result under secant in the table of trigonometric functions.   Read the corresponding angle which will be angle A of triangle ABC.

7.   From angle A of triangle ABC subtract 45° --- which is angle A of triangle ADC.   The remaining angle will be the degree of the elbow which when buttwelded against a 90° elbow as shown in the illustration will result in the given offset.

In this example problem:

Side AC = 1.414 times 9 = 12.726

$$\frac{AC}{AB} = \frac{12.726}{1.5} = 8.41 \quad \text{(Look up under secant.)}$$

From the table of trigonometric functions the angle whose secant is 8.41 is 83°-10'.

83°-10' minus 45° = 38°-10', the angle of the elbow.

Special Offsets
Example problem # 4

Find angle of fitting to use with 90° elbow to make a given offset when the amount of offset is more than the radius of the elbow.

Explanation:

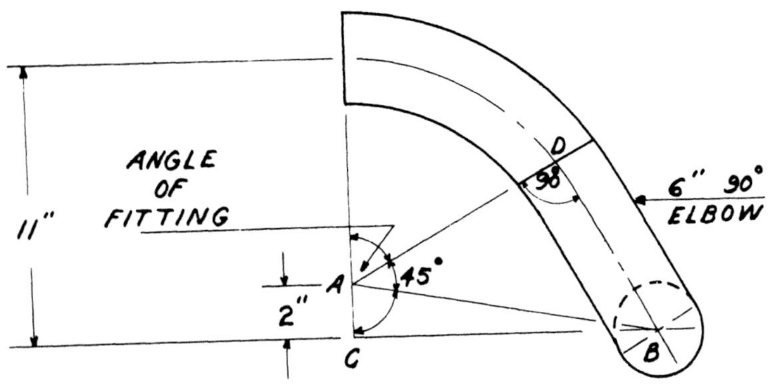

Figure 36

1. Set up problem as shown by drawing radius lines and connecting lines to form triangles CAB and DAB.

2. Side AD and side DB are the radius of the elbows, and since they are of equal length, the acute angles of triangle DAB each contain 45°. (An acute angle is an angle less than 90°.)

3. Find side AB by multiplying side AD by 1.414. (Hypotenuse = opposite side times cosecant.)

4. Find angle CAB. Divide side AB by side AC and look up result under "secant" in table of trigonometric functions. (Hypotenuse divided by adjacent equals secant.) Side AC is found by subtracting the radius of the elbow from the offset.

5. Add angle DAB (45°) to angle CAB and subtract the result from 180° to find the angle of the fitting.

In this example problem:

1.   Side AB = 9 times 1.414 = 12.726 (Opposite side times cosecant of 45°.)

Side AC = 11" - 9 = 2" (Offset minus radius of elbow.)

2.   Angle CAB = 12.726 divided by 2 = 6.363 (Hypotenuse divided by adjacent side.)

3.   From the table of trigonometric functions, the degree that has 6.363 for its secant is 80°-58'.

4.   80°-58' plus 45° = 125°-58'.

5.   180° minus 125°-58' = 54°-02' which is the angle of the fitting that will give 11" of offset when welded against a 90° elbow as shown in example problem # 4.

ORDINATES FOR PIPE RISERS

Ordinates are drop down distances required for the pipe saddle to fit the contour of the outside surface of the run pipe or header. (See figure 37). They are measured from a wraparound line on the riser.

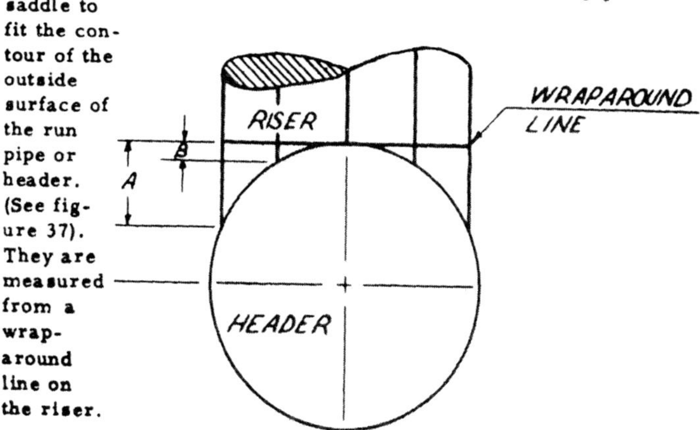

Figure 37

These lengths are
marked on ordinate lines
that are equally spaced
around the circumfer-
ence of the riser pipe.
Four, eight, or six-
teen ordinate lines are
laid off, depending upon
the pipe size. Figure
38 shows eight ordinate
lines and the wrap-
around line scribed on
a length of pipe.

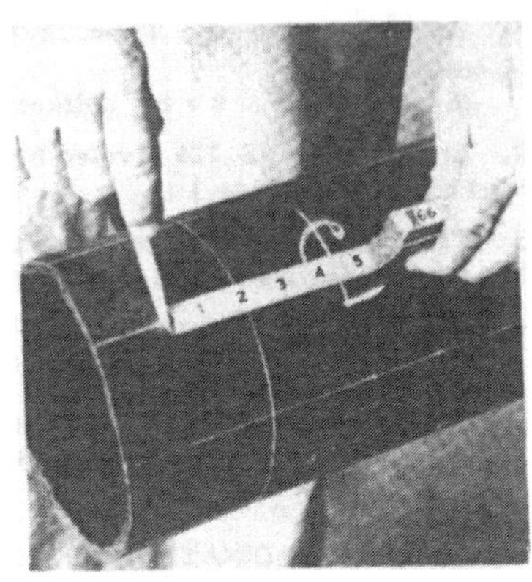

All the ordinates
listed in the tables
in this section are
for "saddle on" type
risers. This means
that the inside dia-
meter of the riser
or branch pipe is
fitted to the out-
side surface of the
header. This type
of fit up allows
the saddle to be
laid off, cut and
placed on the header
in its proper location.
It is then used as a
template to mark the
cutting line for the
opening in the header.
(Figure 39)

42

It is good practice to have the burner "leave the line on the pipe"
to eliminate the possibility of cutting the hole in the header too
large.

The ordinates listed in the following tables are for size to
size and reducing size intersections. For example, look in the
first table and find the ordinates for a 6" saddle on a 6" header.
The header size is read across horizontally and the riser size
is read downward from the top of the table. The ordinates are
found at the crossing point. They are 1 15/16" for the number
2 ordinate and 3/4" for the number 1 ordinate. The 0 ordinate
is in the throat of the saddle so it has no length and is not listed.

In the same manner, the ordinates for a 6" riser on a 12"
header would be 3/4" for the number 2 or longest ordinate and
3/8" for the number 1 ordinate.

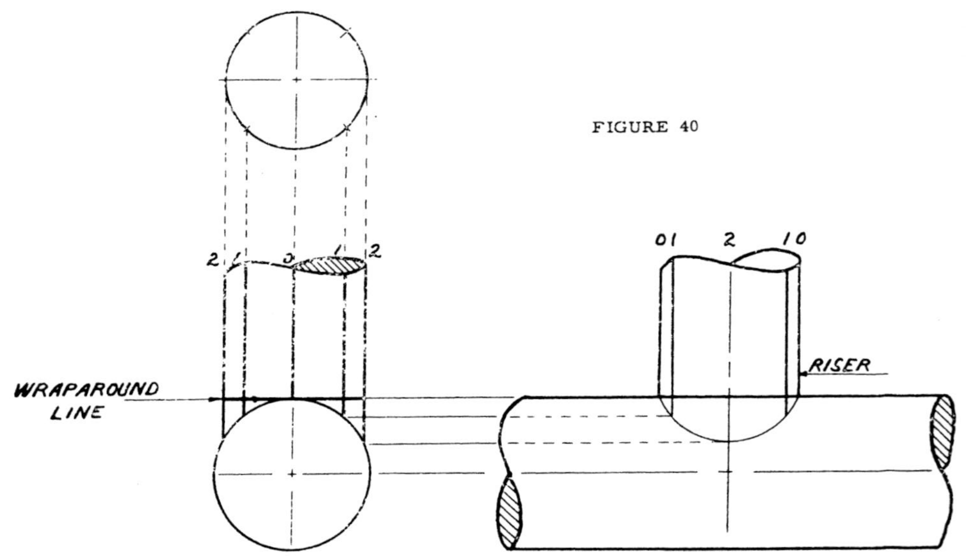

FIGURE 40
WRAPAROUND LINE
RISER

EIGHT ORDINATES FOR 90° INTERSECTIONS
(Standard Wt. Pipe)

PIPE SIZE (Header)

Pipe Size (Riser)	4	6	8	10	12	14	16	18	20	22	24	Ordinate* Number
4"	$\frac{1}{2}$	$\frac{5}{16}$	$\frac{1}{4}$	$\frac{3}{16}$	$\frac{1}{8}$	$\frac{1}{8}$	$\frac{1}{8}$	$\frac{1}{16}$	$\frac{1}{16}$	$\frac{1}{16}$	$\frac{1}{16}$	1
	$1\frac{1}{4}$	$\frac{11}{16}$	$\frac{1}{2}$	$\frac{3}{8}$	$\frac{5}{16}$	$\frac{1}{4}$	$\frac{1}{4}$	$\frac{3}{16}$	$\frac{3}{16}$	$\frac{3}{16}$	$\frac{1}{8}$	2
6"		$\frac{3}{4}$	$\frac{9}{16}$	$\frac{2}{16}$	$\frac{3}{8}$	$\frac{3}{8}$	$\frac{5}{16}$	$\frac{1}{4}$	$\frac{1}{4}$	$\frac{3}{16}$	$\frac{3}{16}$	1
		$1\frac{15}{16}$	$1\frac{1}{4}$	$\frac{15}{16}$	$\frac{3}{4}$	$\frac{11}{16}$	$\frac{5}{8}$	$\frac{1}{2}$	$\frac{1}{2}$	$\frac{7}{16}$	$\frac{3}{8}$	2
8"			$1\frac{1}{16}$	$\frac{13}{16}$	$\frac{11}{16}$	$\frac{5}{8}$	$\frac{9}{16}$	$\frac{1}{2}$	$\frac{7}{16}$	$\frac{3}{8}$	$\frac{3}{8}$	1
			$2\frac{11}{16}$	$1\frac{3}{4}$	$1\frac{2}{16}$	$1\frac{1}{4}$	$1\frac{1}{16}$	$\frac{15}{16}$	$\frac{2}{8}$	$\frac{3}{4}$	$\frac{2}{4}$	2
10"				$1\frac{1}{4}$	$1\frac{1}{16}$	1	$\frac{2}{8}$	$\frac{3}{4}$	$\frac{11}{16}$	$\frac{5}{8}$	$\frac{9}{16}$	1
				$3\frac{7}{16}$	$2\frac{7}{16}$	$2\frac{1}{8}$	$1\frac{3}{4}$	$2\frac{9}{16}$	$1\frac{3}{8}$	$1\frac{1}{8}$	$1\frac{1}{8}$	2

* Ordinate number zero is not listed because it has no length.

SIXTEEN ORDINATES FOR 90° INTERSECTIONS (STD. WT. PIPE)

PIPE SIZE (Header)

PIPE SIZE (RISER)	12	14	16	18	20	22	24	Ordinate* Number
12	7/16	3/8	3/8	5/16	5/16	1/4	1/4	1
	1 5/8	1 9/16	1 3/8	1 1/16	15/16	13/16	7/8	2
	3 1/4	2 3/4	2 5/8	1 15/16	1 13/16	1 1/2	1 3/8	3
	4 1/4	3 7/16	3 1/4	2 5/16	2	1 13/16	1 5/8	4
14		7/16	3/8	3/8	5/16	1/4	1/4	1
		1 7/8	1 1/2	1 15/16	1 1/8	1 1/16	15/16	2
		3 9/16	2 13/16	2 3/8	2 1/16	1 7/8	1 5/8	3
		4 3/4	3 1/2	2 7/8	2 1/2	2 3/16	2	4
16			9/16	1/2	7/16	3/8	3/8	1
			2 1/16	1 13/16	1 9/16	1 3/8	1 1/4	2
			4 3/16	3 3/8	2 7/8	2 9/16	2 1/4	3
			5 9/16	4 1/16	3 1/2	3 1/16	2 11/16	4
18				5/8	9/16	1/2	7/16	1
				2 5/16	2 1/16	1 13/16	1 5/8	2
				4 13/16	3 15/16	3 3/8	3	3
				6 7/16	4 15/16	4 3/16	3 5/8	4
20					11/16	5/8	9/16	1
					2 11/16	2 5/16	2 1/8	2
					5 7/16	4 1/2	3 15/16	3
					6 7/8	5 11/16	4 7/8	4
22						3/4	11/16	1
						2 7/8	2 5/8	2
						6	5 1/16	3
						8 1/8	6 7/16	4
24							7/8	1
							3 1/8	2
							6 5/8	3
							8 1/2	4

*Ordinate number zero
is not listed because
it has no length.

46

EIGHT ORDINATES FOR 90° INTERSECTIONS
(Extra Heavy Pipe)

PIPE SIZE (Header)

Pipe Size (Riser)	4	6	8	10	12	14	16	18	20	22	24	Ordinate* Number
4"	$\frac{7}{16}$	$\frac{5}{16}$	$\frac{1}{4}$	$\frac{3}{16}$	$\frac{3}{16}$	$\frac{1}{8}$	$\frac{1}{8}$	$\frac{1}{8}$	$\frac{1}{8}$	$\frac{1}{8}$	$\frac{1}{16}$	1
	$1\frac{1}{16}$	$\frac{5}{8}$	$\frac{7}{16}$	$\frac{3}{8}$	$\frac{5}{16}$	$\frac{5}{16}$	$\frac{1}{4}$	$\frac{1}{4}$	$\frac{3}{16}$	$\frac{3}{16}$	$\frac{3}{16}$	2
6"		$\frac{11}{16}$	$\frac{1}{2}$	$\frac{3}{8}$	$\frac{3}{8}$	$\frac{5}{16}$	$\frac{1}{4}$	$\frac{1}{4}$	$\frac{1}{4}$	$\frac{3}{16}$	$\frac{3}{16}$	1
		$1\frac{11}{16}$	$1\frac{1}{8}$	$\frac{13}{16}$	$\frac{11}{16}$	$\frac{5}{8}$	$\frac{9}{16}$	$\frac{1}{2}$	$\frac{7}{16}$	$\frac{3}{8}$	$\frac{3}{8}$	2
8"			$\frac{15}{16}$	$\frac{3}{4}$	$\frac{5}{8}$	$\frac{9}{16}$	$\frac{1}{2}$	$\frac{7}{16}$	$\frac{3}{8}$	$\frac{3}{8}$	$\frac{5}{16}$	1
			$2\frac{5}{16}$	$1\frac{5}{8}$	$1\frac{1}{4}$	$1\frac{1}{8}$	1	$\frac{7}{8}$	$\frac{3}{4}$	$\frac{11}{16}$	$\frac{5}{8}$	2
10"				$1\frac{1}{4}$	1	$\frac{15}{16}$	$\frac{13}{16}$	$\frac{11}{16}$	$\frac{5}{8}$	$\frac{9}{16}$	$\frac{1}{2}$	1
				$3\frac{1}{8}$	$2\frac{1}{4}$	2	$1\frac{11}{16}$	$1\frac{7}{16}$	$1\frac{5}{16}$	$1\frac{1}{8}$	$1\frac{1}{16}$	2

* Ordinate number zero is not listed because it has no length.

SIXTEEN ORDINATES FOR 90° INTERSECTIONS (Extra Heavy Pipe)

| Pipe Size (Riser) | PIPE SIZE (Header) | | | | | | | Ordinate* Number |
	12	14	16	18	20	22	24	
	7/16	3/8	5/16	1/4	1/4	1/4	3/16	1
	1 9/16	1 3/8	1 3/16	1 1/8	15/16	13/16	3/4	2
12	3 1/16	2 9/16	2 1/8	1 13/16	1 5/8	1 7/16	1 5/16	3
	3 11/16	3 3/16	2 1/2	2 3/16	1 15/16	1 11/16	1 9/16	4
		7/16	7/16	3/8	5/16	5/16	1/4	1
		1 3/4	1 1/2	1 1/4	1 1/8	1	15/16	2
14		3 7/16	2 3/4	2 5/16	2	1 13/16	1 5/8	3
		4 7/16	3 3/8	2 13/16	2 7/16	2 1/8	1 15/16	4
			9/16	1/2	7/16	3/8	3/8	1
			2	1 3/4	1 9/16	1 3/8	1 1/4	2
16			4	3 1/4	2 13/16	2 1/2	2 3/16	3
			5 1/4	4 1/8	3 3/8	3	2 5/8	4
				5/8	9/16	1/2	7/16	1
				2 5/16	2	1 11/16	1 5/8	2
18				4 5/8	3 13/16	3 5/16	3	3
				6 1/16	4 3/4	4	3 9/16	4
					11/16	5/8	9/16	1
					2 5/8	2 5/16	2 1/16	2
20					5 1/4	4 3/8	3 13/16	3
					6 7/8	6	4 11/16	4
						3/4	11/16	1
						2 7/8	2 9/16	2
22						5 13/16	4 15/16	3
						7 3/4	6 3/16	4
							7/8	1
							3 3/16	2
24							6 7/16	3
							8 9/16	4

* Ordinate number zero is not listed because it has no length.

48

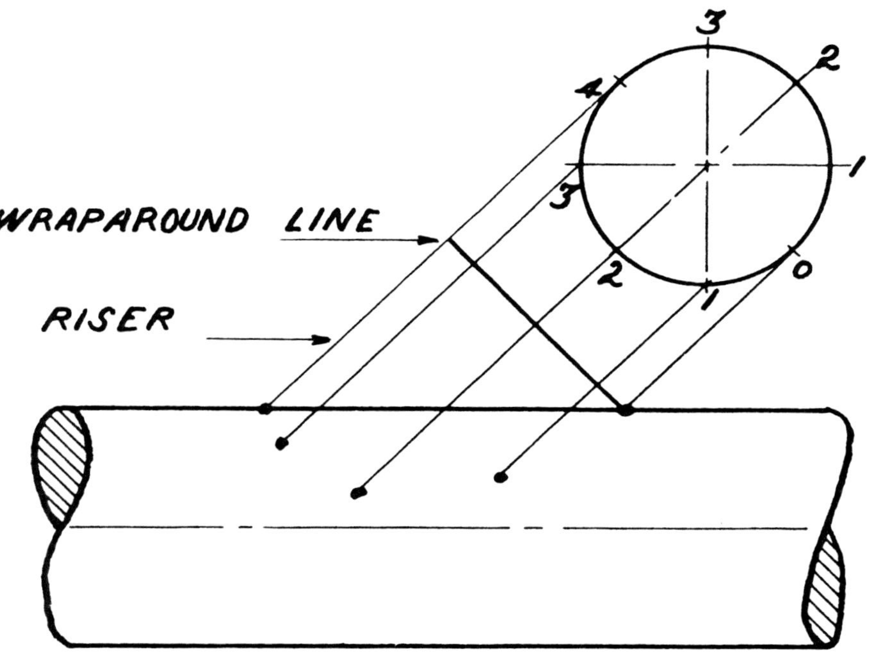

3
4
2
3
1
2
0
1
WRAPAROUND LINE
RISER

EIGHT ORDINATES FOR 45° PIPE INTERSECTIONS
(Standard Weight Pipe)

Pipe Size (Riser)	PIPE SIZE (Header)											Ordinate Number
	4	6	8	10	12	14	16	18	20	22	24	
4	1 1/4	1	15/16	13/16	13/16	3/4	3/4	11/16	11/16	11/16	11/16	1
	3 11/16	2 15/16	2 11/16	2 1/2	2 7/16	2 3/8	2 5/16	2 5/16	2 5/16	2 1/4	2 1/4	2
	4 1/8	3 13/16	3 3/4	3 11/16	3 5/8	3 9/16	3 9/16	3 1/2	3 9/16	3 1/2	3 1/2	3
	4	4	4	4	4	4	4	4	4	4	4	4
6		1 15/16	1 11/16	1 1/2	1 3/8	1 3/8	1 1/4	1 1/4	1 1/4	1 3/16	1 1/8	1
		5 3/4	4 11/16	4 1/4	4 1/16	3 15/16	3 3/16	3 3/4	3 11/16	3 9/16	3 1/2	2
		6 3/16	5 2/8	5 3/4	5 5/8	5 9/16	5 1/2	5 7/16	5 2/16	5 2/16	5 3/8	3
		6	6	6	6	6	6	6	6	6	6	4
8			2 11/16	2 5/16	2 1/16	2	1 15/16	1 11/16	1 3/4	1 11/16	1 5/8	1
			7 13/16	6 2/8	6	5 3/4	5 1/2	5 1/4	5 3/16	5 1/16	4 15/16	2
			8 5/16	7 15/16	7 3/4	7 5/8	7 9/16	7 1/2	7 3/8	7 5/16	7 5/16	3
			8	8	8	8	8	8	8	8	8	4
10				3 3/16	2 5/8	2 11/16	2 5/8	2 7/8	2 3/8	2 5/16	2 3/16	1
				9 13/16	8 7/16	7 15/16	7 2/8	7 1/8	6 7/8	6 9/16	6 9/16	2
				10 1/4	10	9 7/8	9 11/16	9 9/16	9 7/16	9 3/8	9 1/4	3
				10	10	10	10	10	10	10	10	4

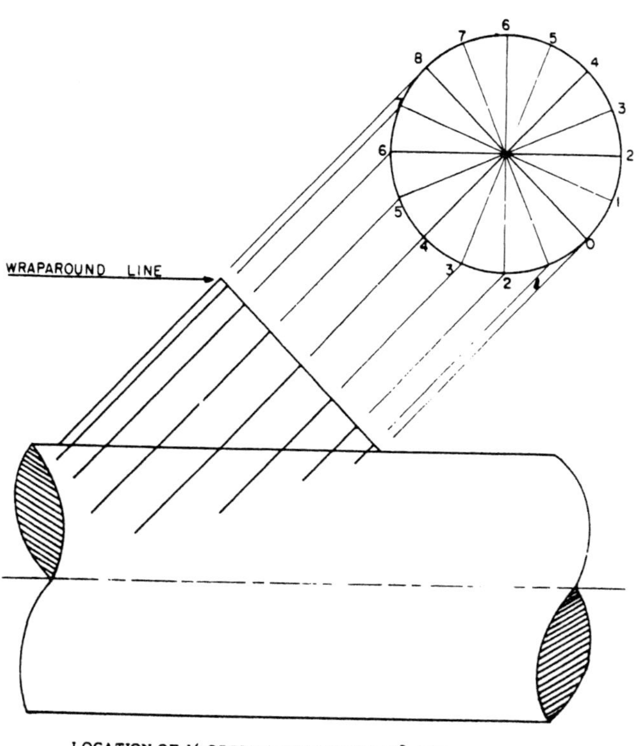

LOCATION OF 16 ORDINATE LINES FOR 45° INTERSECTIONS

SIXTEEN ORDINATES FOR 45° PIPE INTERSECTIONS
(Standard Weight Pipe)

Pipe Size (Riser)	12	16	20	24	Ordinate Number
12	1 1/16	15/16	13/16	3/4	1
	4 1/16	3 7/16	3 1/16	2 11/16	2
	8 1/8	6 7/8	6 3/16	5 5/8	3
	11 15/16	9 11/16	8 11/16	8 5/16	4
	12 13/16	11 7/16	10 13/16	10 1/4	5
	12 1/2	11 15/16	11 9/16	11 5/16	6
	12 1/8	12	11 15/16	11 13/16	7
	12	12	12	12	8
16		1 3/8	1 3/16	1 1/8	1
		5 3/16	4 7/16	4 1/16	2
		10 11/16	8 13/16	7 15/16	3
		15 1/2	12 5/8	11 7/16	4
		16 1/2	14 5/8	13 3/4	5
		15 15/16	15 3/16	14 11/16	6
		15 7/16	15 1/4	15 3/16	7
		15 1/4	15 1/4	15 1/4	8
20			1 11/16	1 1/2	1
			6 5/8	5 13/16	2
			13 5/8	11 1/2	3
			19 7/8	16 7/16	4
			21	18 7/8	5
			20 1/4	19 7/16	6
			19 1/2	19 5/16	7
			19 1/4	19 1/4	8
24				2 1/8	1
				8	2
				16 9/16	3
				24 3/8	4
				25 1/2	5
				24 7/16	6
				23 9/16	7
				23 1/4	8

EIGHT ORDINATES FOR 45° PIPE INTERSECTIONS
(Extra Heavy Pipe)

PIPE SIZE (Header)

Pipe Size (Riser)	4	6	8	10	12	16	20	24	Ordinate Number
4	1 3/16	15/16	7/8	13/16	3/4	11/16	11/16	5/8	1
	3 3/8	2 3/4	2 1/2	2 3/8	2 5/16	2 3/16	2 3/16	2 1/8	2
	3 7/8	3 5/8	3 9/16	3 1/2	3 1/2	3 7/16	3 3/8	3 5/16	3
	3 13/16	3 13/16	3 13/16	3 13/16	3 13/16	3 13/16	3 13/16	3 13/16	4
6		1 13/16	1 9/16	1 3/8	1 5/16	1 3/16	1/4	1 1/8	1
		5 1/4	4 7/16	4	3 11/16	3 5/8	3 1/2	3 3/8	2
		5 13/16	5 5/8	5 7/16	5 3/8	5 1/4	5 3/16	5 3/16	3
		5 3/4	5 3/4	5 3/4	5 3/4	5 3/4	5 3/4	5 3/4	4
8			2 7/16	2 3/16	1 15/16	1 11/16	1 5/8	1 9/16	1
			7 1/16	6 1/16	5 5/8	5 3/16	4 7/8	4 11/16	2
			7 11/16	7 5/8	7 3/8	7 3/16	7 1/16	6 15/16	3
			7 5/8	7 5/8	7 5/8	7 5/8	7 5/8	7 5/8	4
10				3 3/16	2 11/16	2 1/2	2 5/16	2 1/8	1
				9 5/16	8 1/16	7 3/16	6 11/16	6 5/16	2
				10 1/8	9 11/16	9 7/16	9 3/16	9	3
				9 3/4	9 3/4	9 3/4	9 3/4	9 3/4	4

SIXTEEN ORDINATES FOR 45° PIPE INTERSECTIONS
(Extra Heavy Pipe)

Pipe Size (Riser)	PIPE SIZE (Header)				Ordinate Number
	12	16	20	24	
12	1	7/8	13/16	11/16	1
	3 7/8	3 3/8	3	2 3/4	2
	7 7/8	6 5/8	5 7/8	5 7/16	3
	11 1/16	9 3/8	5 9/16	8	4
	12 3/8	11 1/8	10 3/8	9 15/16	5
	12 3/16	11 11/16	11 5/16	11 1/16	6
	11 7/8	11 3/4	11 5/8	11 9/16	7
	11 3/4	11 3/4	11 3/4	11 3/4	8
16		1 5/16	1 1/8	1 1/16	1
		5	4 3/8	3 15/16	2
		10 1/4	8 9/16	7 11/16	3
		14 7/8	12 1/4	11 3/16	4
		16	14 5/16	13 7/16	5
		15 5/8	14 15/16	14 9/16	6
		15 3/16	15	14 7/8	7
		15	15	15	8
20			1 11/16	1 1/2	1
			6 7/16	5 11/16	1
			13 1/4	11 1/4	3
			19 3/16	16 1/16	4
			20 1/2	18 1/2	5
			19 7/8	19 1/8	6
			19 1/4	19 1/16	7
			19	19	8
24				2 1/16	1
				7 7/8	2
				16 3/16	3
				23 5/8	4
				25	5
				24 1/8	6
				23 5/16	7
				23	8

SIXTEEN ORDINATES FOR 90° ECCENTRIC PIPE RISERS
(Standard Weight Pipe)

PIPE SIZE (Header)

Pipe Size (Riser)	4	6	8	10	12	16	20	24	
	1/4	0	1/8	3/8	3/4	1 7/16	2 1/2	3 5/8	0
	3/16	1/16	1/8	7/16	3/16	1 9/16	2 5/8	3 3/4	1
	1/16	1/16	3/16	9/16	1	1 3/4	2 15/16	4 1/8	2
	0	1/8	7/16	7/8	1 5/16	2 1/4	3 7/16	4 3/4	3
3	1/16	3/8	13/16	1 1/4	1 7/8	2 11/16	4 3/16	5 9/16	4
	1/4	11/16	1 1/4	1 7/8	2 1/2	3 5/8	5 1/16	6 9/16	5
	5/8	1 3/16	2	2 1/2	3 1/4	4 1/2	6 1/16	7 5/8	6
	1	1 11/16	2 1/4	3 1/8	3 13/16	5 1/4	6 7/8	8 9/16	7
	1 1/2	1 7/8	2 11/16	3 3/8	4 3/16	5 9/16	7 1/4	9	8
		3/16	0	1/16	5/16	7/8	1 11/16	2 3/4	0
		1/8	0	1/8	3/8	15/16	1 13/16	2 7/8	1
		1/16	1/16	1/4	9/16	1 3/16	2 1/8	3 1/4	2
		0	3/16	1/2	7/8	1 5/8	2 11/16	3 7/8	3
4		1/8	7/16	15/16	1 3/8	2 5/16	3 1/2	4 13/16	4
		1/2	1	1 1/2	2 1/8	3 3/16	4 9/16	6	5
		1	1 5/8	2 5/16	3	4 3/16	5 11/16	7 1/4	6
		1 5/8	2 1/4	3 1/16	3 13/16	5 1/8	6 13/16	8 7/16	7
		1 15/16	2 9/16	3 3/8	4 3/16	5 9/16	7 1/8	9	8
			1/2	1/16	0	1/8	11/16	1 3/8	0
			3/8	1/16	0	3/16	3/4	1 1/2	1
			1/4	0	1/16	1/16	1 1/8	1 7/8	2
			0	1/16	1/4	3/4	1 5/8	2 5/8	3
6			1/8	3/8	3/4	1 7/16	2 1/2	3 5/8	4
			9/16	1	1 1/2	2 7/16	3 3/4	5	5
			1 3/8	1 7/8	2 9/16	3 11/16	5 1/8	6 5/8	6
			2 3/16	2 7/8	3 5/8	4 15/16	6 9/16	8 1/4	7
			2 11/16	3 3/8	4 3/16	5 9/16	7 1/4	9	8

SIXTEEN ORDINATES FOR 90º ECCENTRIC PIPE RISERS
(Standard Weight Pipe)
(Continued)

Pipe
Size
(Riser)

PIPE SIZE (Header)

Pipe Size (Riser)	10	12	16	
	7/8	5/16	0	0
	11/16	3/16	0	1
	5/16	1/8	0	2
	0	0	1/4	3
	1/16	5/16	7/8	4
	5/8	1 1/16	1 7/8	5
	1 9/16	2 3/16	3 1/4	6
	2 3/4	3 1/2	4 3/4	7
	3 3/8	4 3/16	5 9/16	8
		1 3/8	3/8	0
		1 1/8	1/4	1
		1/2	1/16	2
		1/16	0	3
		1/16	7/16	4
10		11/16	1 3/8	5
		1 7/8	2 7/8	6
		3 3/8	4 9/16	7
		4 3/16	5 9/16	8
			1 5/16	0
			15/16	1
			7/16	2
			0	3
12			1/8	4
			1	5
			2 9/16	6
			4 7/16	7
			5 9/16	8

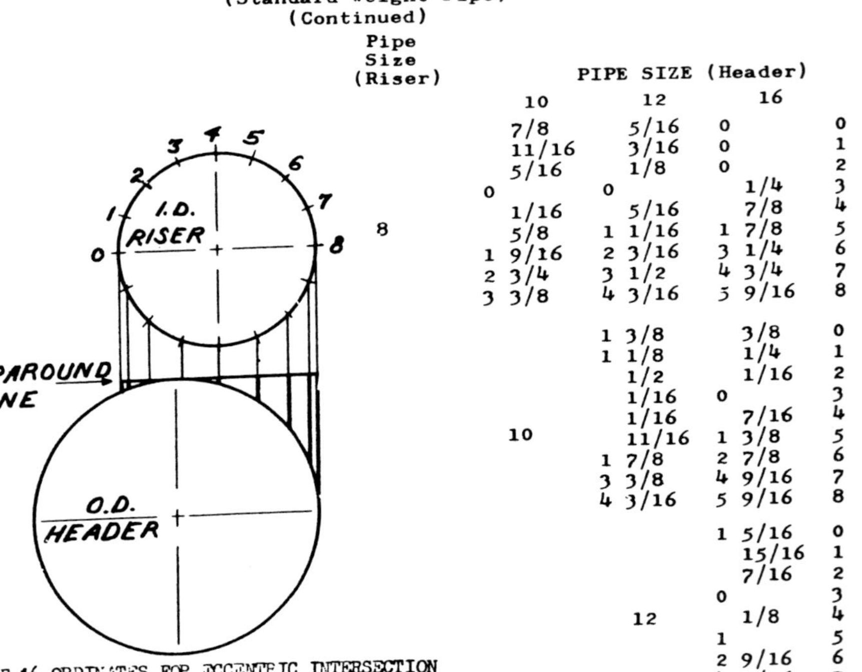

LOCATION OF 16 ORDINATES FOR ECCENTRIC INTERSECTION

SIXTEEN ORDINATES FOR CONCENTRIC RISER
ON BACK OF ELBOW (Standard Wt. Pipe)

Pipe Size (Riser)	PIPE SIZE (Header)						Ordinate Number
	4	6	8	10	12	16	
4	4 5/8	7 7/8	11 1/8	14 5/16	17 5/8	24 3/4	0
	5 1/4	8 1/16	11 5/16	14 1/2	17 7/8	25	1
	5 3/4	8 3/4	11 15/16	15 1/16	18 3/8	25 1/2	2
	6 15/16	9 5/8	12 13/16	15 3/4	19 1/8	26 5/16	3
	8 5/16	10 5/8	13 11/16	16 3/4	20 1/16	27 3/16	4
	9 5/16	11 1/2	14 9/16	17 5/8	20 15/16	28 1/16	5
	9 3/4	12 1/16	15 3/16	18 1/4	21 1/2	28 3/4	6
	9 15/16	12 3/8	15 1/2	18 11/16	21 13/16	29 1/4	7
	10	12 1/2	15 5/8	18 3/4	22 1/16	29 5/16	8
6		7 1/4	10 3/8	13 1/2	16 3/4	23 13/16	0
		7 5/8	10 3/4	13 13/16	17 1/8	24 3/16	1
		8 3/4	11 11/16	14 13/16	18	25 1/16	2
		10 9/16	13 3/16	16 1/8	19 5/16	26 3/8	3
		12 13/16	14 7/8	17 9/16	20 11/16	27 3/4	4
		14 1/4	16 5/16	18 7/8	22	29 1/16	5
		14 15/16	17	19 3/4	22 7/8	30	6
		15 3/16	17 7/16	20 1/4	23 1/2	30 9/16	7
		15 1/4	17 9/16	20 7/16	23 5/8	30 7/8	8
8			9 3/4	12 7/8	16 1/16	23 1/16	0
			10 1/4	13 5/16	16 1/2	23 1/2	1
			11 13/16	14 5/8	17 3/4	24 3/4	2
			14 3/8	16 11/16	19 11/16	26 5/8	3
			17 9/16	19 1/16	21 15/16	28 1/2	4
			19 5/8	20 13/16	23 1/2	30 7/16	5
			20 7/16	21 15/16	24 11/16	31 7/16	6
			20 11/16	22 1/2	25 5/16	32 1/4	7
			20 3/4	22 3/4	25 1/2	32 1/2	8

SIXTEEN ORDINATES FOR CONCENTRIC RISER ON BACK OF ELBOW (Standard Wt. Pipe)
(Continued)

Pipe Size (Riser)	PIPE SIZE (Header)			Ordinate Number
	10	12	16	
10	12 1/4	15 3/8	22 1/4	0
	12 7/8	16 7/16	22 13/16	1
	14 15/16	17 3/4	24 7/16	2
	18	20 1/2	26 15/16	3
	22 1/16	23 15/16	29 9/16	4
	24 9/16	25 15/16	31 13/16	5
	25 5/8	27 1/8	33 1/2	6
	26	27 3/4	34 1/4	7
	26 1/8	28	34 1/2	8
12		14 13/16	21 9/16	0
		15 9/16	22 1/4	1
		17 7/8	24 5/16	2
		21 13/16	27 11/16	3
		26 15/16	31 1/4	4
		30 1/16	32 1/8	5
		31 1/4	36 3/8	6
		31 9/16	36 11/16	7
		31 3/4	36 15/16	8
16			20 1/2	0
			21 9/16	1
			24 9/16	2
			30 1/16	3
			36 15/16	4
			40 11/16	5
			42 1/2	6
			43 1/16	7
			43 3/16	8

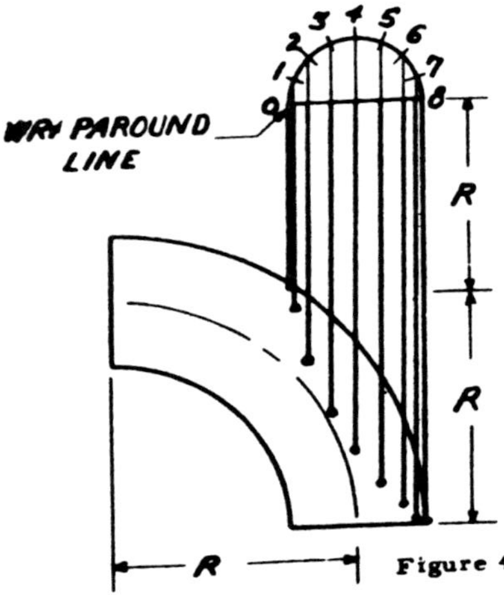

Figure 44

Note: These ordinates are measured from a wrap-around line that is the length of two radii from the end of the elbow.

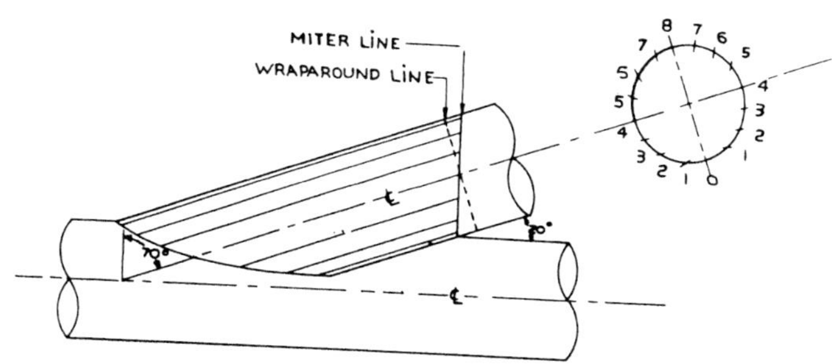

EXAMPLE: LAY OUT 20° RISER

1. LAY OFF 16 ORDINATE LINES ON THE RISER PIPE

2. LAY OFF MITER LINE ON THE RISER PIPE EQUAL TO THE COMPLEMENT OF
    THE DEGREE OF INTERSECTION. COMPLEMENT IS THE DEGREE OF INTER-
    SECTION SUBTRACTED FROM 90°.   IN THIS EXAMPLE THE MITER LINE WILL
    BE 70° (90° MINUS 20° EQUALS 70°). NOTE: USE I.D. OF RISER PIPE
    WHEN FIGURING CUT BACK FOR THE MITER LINE.

3. FIND THE 90° CUT BACK LENGTHS FROM THE TABLES IN THIS BOOK AND
    MULTIPLY EACH ONE BY THE COSECANT OF THE DEGREE OF INTERSECTION.
    IN THIS CASE IT WILL BE COSECANT OF 20° TIMES EACH OF THE 90°
    CUT BACK LENGTHS. THIS WILL GIVE YOU THE ORDINATE LENGTHS FOR
    A 20° INTERSECTION.

4. LAY OFF THESE ORDINATE LENGTHS ON THE ORDINATE LINES. MEASURE
    THEM FROM THE <u>MITER</u> LINE, NOT THE WRAPAROUND LINE. BE SURE TO
    LET THE SIDE CENTERS OF THE MITER (# 4 ORDINATE) BE THE SIDE
    CENTERS OF THE RISER.

5. CONNECT THE POINTS INTO A CURVING LINE, SAME AS FOR ANY OTHER
    INTERSECTION. MAKE RADIAL CUT.

    NOTE: FOR A MORE ACCURATE FIT, USE 16 ORDINATE LINES.

HOW TO LAY OUT 45° ECCENTRIC PIPE RISER

1. Lay out a 45 degree miter cutting line on the pipe. Use I. D. of pipe when figuring cut back length of the miter.

2. Find the 90° eccentric ordinates for the intersection on pages 54 or 55. Multiply each ordinate length times 1.41. This will give you the ordinates for the 45° eccentric intersection.

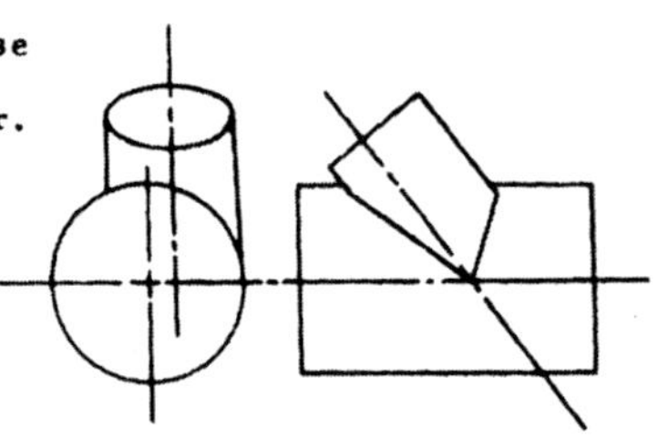

Figure 47

3. Lay out 16 ordinate lines on the riser pipe at the 45° miter line and number them 0 thru 8 (figure 48).

4. Mark the ordinate lengths on the ordinate lines, measuring from the 45° miter line.

5. Connect the points into a curving line, the same as for any saddle type riser.

NOTE: The side of the riser that has #8 ordinate on it will be tangent to the header (see figure 48).

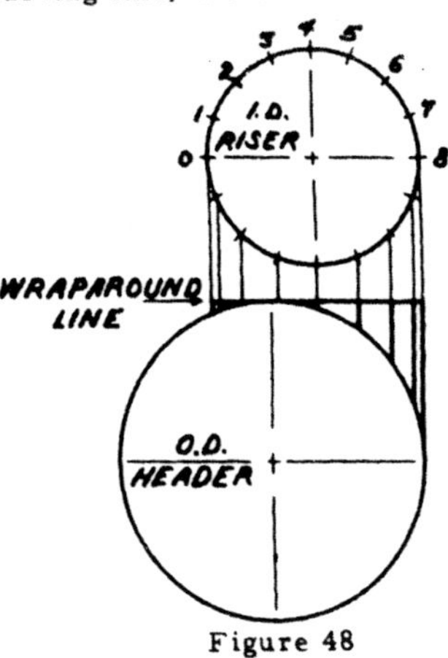

Figure 48

DIMENSIONS OF H BEAM PIPE SLIDERS

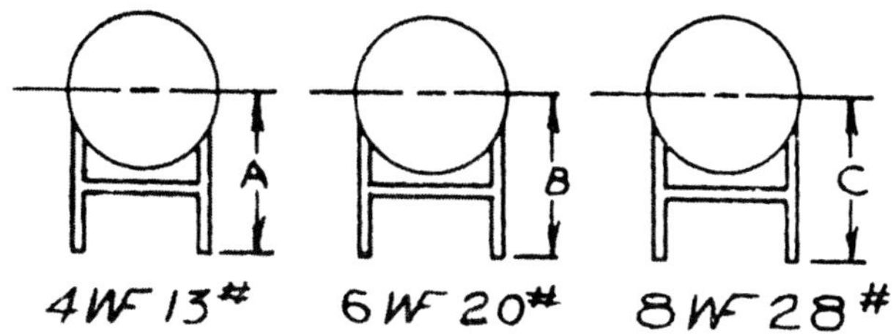

Dimensions in Inches

Pipe Size	A	B	C
4	5 1/2		
6	6 7/8		
8	8		
10	9 1/8		
12		11 3/4	
14		12 1/2	
16		13 1/2	
18			14 3/4
20			15 13/16
24			18

DIMENSIONS OF SEAMLESS STEEL PIPE

PIPE SIZE	O.D.	SCH. 10	SCH. 20	SCH. 30	STD. WT.	SCH. 40	SCH. 60	EX. STG.	SCH. 80	SCH. 100	SCH. 120	SCH. 140	SCH. 160	XX STG.
1/8	.405				.068	.068		.095	.095					
1/4	.540				.088	.088		.119	.119					
3/8	.675				.091	.091		.126	.126					
1/2	.840				.109	.109		.147	.147				.187	.294
3/4	1.050				.113	.113		.154	.154				.218	.308
1	1.315				.133	.133		.179	.179				.250	.358
1 1/4	1.660				.140	.140		.191	.191				.250	.382
1 1/2	1.900				.145	.145		.200	.200				.281	.400
2	2.375				.154	.154		.218	.218				.343	.436
2 1/2	2.875				.203	.203		.276	.276				.375	.552
3	3.500				.216	.216		.300	.300				.438	.600
3 1/2	4.000				.226	.226		.318	.318				....	.636
4	4.500				.237	.237		.337	.337		.438		.531	.674
5	5.563				.258	.258		.375	.375		.500		.625	.750
6	6.625				.280	.280		.432	.432		.562		.718	.864
8	8.625		.250	.277	.322	.322	.406	.500	.500	.593	.718	.812	.906	.875
10	10.75		.250	.307	.365	.365	.500	.500	.593	.718	.843	1.000	1.125	....
12	12.75		.250	.330	.375	.406	.562	.500	.687	.843	1.000	1.125	1.312	....
14 OD	14.00	.250	.312	.375	.375	.438	.593	.500	.750	.937	1.093	1.250	1.406	....
16 OD	16.00	.250	.312	.375	.375	.500	.656	.500	.843	1.031	1.218	1.438	1.593	....
18 OD	18.00	.250	.312	.438	.375	.562	.750	.500	.937	1.156	1.375	1.562	1.781	....
20 OD	20.00	.250	.375	.500	.375	.593	.812	.500	1.031	1.281	1.500	1.750	1.968	....
24 OD	24.00	.250	.375	.562	.375	.687	.968	.500	1.218	1.531	1.812	2.062	2.343	....
30 OD	30.00	.316	.500	.625	.375	....	....	.500	....	....	....	....	....	....

FACE TO FACE DIMENSIONS OF FLANGED AND WELDING END VALVES

Nom. Pipe Size	Cast Iron		Steel								Nom. Pipe Size
					Welding or Flanged Ends						
	125 lb. Flanged	250 lb. Flanged	150 lb. Flanged	150 lb. Wldg. End	300 lb.	400 lb.	600 lb.	900 lb.	1500 lb.	2500 lb.	
1	—	—	5	5	—	8 ½	8 ½	10	10	12⅛	1
1¼	—	—	5 ½	5 ½	—	9	9	11	11	13 ¾	1¼
1½	—	—	6 ½	6 ½	7 ½	9 ½	9 ½	12	12	15⅛	1½
2	7	8 ½	7	8 ½	8 ½	11 ½	11 ½	14 ½	14 ½	17 ¾	2
2½	7 ½	9 ½	7 ½	9 ½	9 ½	13	13	16 ½	16 ½	20	2½
3	8	11⅛	8	11⅛	11⅛	14	14	15	18 ½	22 ¾	3
3½	8 ½	11⅞	8 ½	—	11⅞	15	15	17	19 ½	—	3½
4	9	12	9	12	12	16	17	18	21 ½	26 ½	4
5	10	15	10	15	15	18	20	22	26 ½	31 ¼	5
6	10 ½	15⅞	10 ½	15⅞	15⅞	19 ½	22	24	27 ¾	36	6
8	11 ½	16 ½	11 ½	16 ½	16 ½	23 ½	26	29	32 ¾	40 ¼	8
10	13	18	13	18	18	26 ½	31	33	39	50	10
12	14	19 ¾	14	19 ¾	19 ¾	30	33	38	44 ½	56	12
14	15	22 ½	15	22 ½	30	32 ½	35	40 ½	49 ½	—	14
16	16	24	16	24	33	35 ½	39	44 ½	54 ½	—	16
18	17	26	17	26	36	38 ½	43	48	60 ½	—	18
20	18	28	18	28	39	41 ½	47	52	65 ½	—	20
24	20	31	20	32	45	48 ½	55	61	76 ½	—	24

FACE TO FACE DIMENSIONS OF FLANGED AND WELDING END GLOBE VALVES

Nom. Pipe Size	Cast Iron		150 lb. Flanged	150 lb. Wldg. End	Steel Welding or Flanged Ends						Nom. Pipe Size
	125 lb. Flanged	250 lb. Flanged			300 lb.	400 lb.	600 lb.	900 lb.	1500 lb.	2500 lb.	
1	—	—	5	5	8	8 ½	8 ½	10	10	12 ½	1
1¼	—	—	5 ½	5 ½	8 ½	9	9	11	11	13 ¾	1¼
1½	—	—	6 ½	6 ½	9	9 ½	9 ½	12	12	15 ½	1½
2	8	10 ½	8	8	10 ½	11 ½	11 ½	14 ½	14 ½	17 ¾	2
2½	8 ½	11 ½	8 ½	8 ½	11 ½	13	13	16 ½	16 ½	20	2½
3	9 ½	12 ½	9 ½	9 ½	12 ½	14	14	15	18 ½	22 ¾	3
3½	10 ½	13 ¼	10 ½	10 ½	13 ¼	15	15	17	19 ½	—	3½
4	11 ½	14	11 ½	11 ½	14	16	17	18	21 ½	26 ½	4
5	13	15 ¾	14	14	15 ¾	18	20	22	26 ½	31 ¼	5
6	14	17 ½	16	16	17 ½	19 ½	22	24	27 ¾	36	6
8	19 ½	21	19 ½	19 ½	22	23 ½	26	29	32 ¾	40 ¼	8
10	24 ½	24 ½	24 ½	24 ½	24 ½	26 ½	31	33	39	50	10
12	27 ½	28	27 ½	27 ½	28	30	33	38	44 ½	56	12

THE FACE TO FACE DIMENSIONS OF FLANGED SWING CHECK VALVES ARE THE SAME AS
GLOBE VALVES EXCEPT THE FOLLOWING:

    5" 150 Lb....13"   1 1/4" 300 Lb....9"
    6" 150 Lb....14"   1 1/2" 300 Lb....9 1/2"
    1" 300 Lb.... 8 1/2"   8"   300 Lb....21"

LENGTH THRU HUB
OF
WELDING NECK FLANGES

PIPE SIZE	150 Lb.	300 Lb.	400 Lb.	600 Lb.	900 Lb.	1500 Lb.	2500 Lb.
1/2	1 7/8	2 1/16	2 1/16	2 1/16	2 3/8	2 3/8	2 7/8
3/4	2 1/16	2 1/4	2 1/4	2 1/4	2 3/4	2 3/4	3 1/8
1	2 3/16	2 7/16	2 7/16	2 7/16	2 7/8	2 7/8	3 1/2
1 1/4	2 1/4	2 9/16	2 5/8	2 5/8	2 7/8	2 7/8	3 3/4
1 1/2	2 7/16	2 11/16	2 3/4	2 3/4	3 1/4	3 1/4	4 3/8
2	2 1/2	2 3/4	2 7/8	2 7/8	4	4	5
2 1/2	2 3/4	3	3 1/8	3 1/8	4 1/8	4 1/8	5 5/8
3	2 3/4	3 1/8	3 1/4	3 1/4	4	4 5/8	6 5/8
3 1/2	2 13/16	3 3/16	3 3/8	3 3/8	....	....	....
4	3	3 3/8	3 1/2	4	4 1/2	4 7/8	7 1/2
6	3 1/2	3 7/8	4 1/16	4 5/8	5 1/2	6 3/4	10 3/4
8	4	4 3/8	4 5/8	5 1/4	6 3/8	8 3/4	12 1/2
10	4	4 5/8	4 7/8	6	7 1/4	10	16 1/2
12	4 1/2	5 1/8	5 3/8	6 1/8	7 7/8	11 1/8	18 1/4
14	5	5 5/8	5 7/8	6 1/2	8 3/8	11 3/4	....
16	5	5 3/4	6	7	8 1/2	12 1/4	....
18	5 1/2	6 1/4	6 1/2	7 1/4	9	12 7/8	....
20	5 11/16	6 3/8	6 5/8	7 1/2	9 3/4	14	....
24	6	6 5/8	6 7/8	8	11 1/2	15	....

ADD 1/4" FOR RAISED FACE TO ALL
FLANGE DIMENSIONS ABOVE 400 LBS.

65

OUTSIDE DIAMETERS OF STEEL FLANGES

Pressure Classes

Pipe Size (Inches)	150 Lb.	300 Lb.	400 Lb.	600 Lb.	900 Lb.	1500 Lb.	2500 Lb.
1/2	3 1/2	3 3/4	-	3 3/4	-	4 3/4	5 1/4
3/4	3 7/8	4 5/8	-	4 5/8	-	5 1/8	5 1/2
1	4 1/4	4 7/8	-	4 7/8	-	5 7/8	6 1/4
1 1/4	4 5/8	5 1/4	-	5 1/4	-	6 1/4	7 1/4
1 1/2	5	6 1/8	-	6 1/8	-	7	8
2	6	6 1/2	-	6 1/2	-	8 1/2	9 1/4
2 1/2	7	7 1/2	-	7 1/2	-	9 5/8	10 1/2
3	7 1/2	8 1/4	-	8 1/4	9 1/2	10 1/2	12
4	9	10	10	10 3/4	11 1/2	12 1/4	14
5	10	11	11	13	13 3/4	14 3/4	16 1/2
6	11	12 1/2	12 1/2	14	15	15 1/2	19
8	13 1/2	15	15	16 1/2	18 1/2	19	21 3/4
10	16	17 1/2	17 1/2	20	21 1/2	23	26 1/2
12	19	20 1/2	20 1/2	22	24	26 1/2	30
14	21	23	23	23 3/4	25 1/4	29 1/2	*
16	23 1/2	25 1/2	25 1/2	27	27 3/4	*	*
18	25	28	28	29 1/4	31	*	*
20	27 1/2	30 1/2	30 1/2	32	33 3/4	*	*
24	32	36	36	37	41	*	*

For 400 Lb., where no dimension is given use 600 Lb.
For 900 Lb., where no dimension is given use 1500 Lb.

66

DIMENSIONS OF GASKETS FOR RAISED FACE FLANGES

PIPE SIZE	INSIDE DIA.	150 Lb.	300 Lb.	400 Lb.	600 Lb.	900 Lb.	1500 Lb.
1/2	7/8	1 7/8	2 1/8	...	2 1/8	...	2 1/2
3/4	1 1/16	2 1/4	2 5/8	...	2 5/8	...	2 3/4
1	1 5/16	2 5/8	2 7/8	...	2 7/8	...	3 1/8
1 1/4	1 5/8	3	3 1/4	...	3 1/4	...	3 1/2
1 1/2	1 7/8	3 3/8	3 3/4	...	3 3/4	...	3 7/8
2	2 3/8	4 1/8	4 3/8	...	4 3/8	...	5 5/8
2 1/2	2 7/8	4 7/8	5 1/8	...	5 1/8	...	6 1/2
3	3 1/2	5 3/8	5 7/8	...	5 7/8	6 5/8	6 7/8
3 1/2	4	6 3/8	6 1/2	...	6 3/8	...	...
4	4 1/2	6 7/8	7 1/8	7	7 5/8	8 1/8	8 1/4
5	5 9/16	7 3/4	8 1/2	8 3/8	9 1/2	9 3/4	10
6	6 5/8	8 3/4	9 7/8	9 3/4	10 1/2	11 3/8	11 1/8
8	8 5/8	11	12 1/8	12	12 5/8	14 1/8	13 7/8
10	10 3/4	13 3/8	14 1/4	14 1/8	15 3/4	17 1/8	17 1/8
12	12 3/4	16 1/8	16 5/8	16 1/2	18	19 5/8	20 1/2
14	14	17 3/4	19 1/8	19	19 3/8	20 1/2	22 3/4
16	16	20 1/4	21 1/4	21 1/8	22 1/4	22 5/8	...
18	18	21 5/8	23 1/2	23 3/8	24 1/8	25 1/8	...
20	20	23 7/8	25 3/4	25 1/2	26 7/8	27 1/2	...
24	24	28 1/4	30 1/2	30 1/4	31 1/8	33	...

LENGTH OF BOLTS FOR PIPE FLANGES

Nominal Pipe Size	150 LB. FLANGES				300 LB. FLANGES				400 LB. FLANGES		
	Number of Bolts	Diam. of Bolts	Length Mach. Bolts	Length Stud Bolts	Number of Bolts	Diam. of Bolts	Length Mach. Bolts	Length Stud Bolts	Number of Bolts	Diam. of Bolts	Length of Stud Bolts
½	4	½	1¾	2¼	4	½	2	2½	4	½	3
¾	4	½	2	2¼	4	⅝	2½	2¾	4	⅝	3¼
1	4	½	2	2½	4	⅝	2½	3	4	⅝	3½
1¼	4	½	2¼	2½	4	⅝	2¾	3	4	⅝	3¾
1½	4	½	2¼	2¾	4	¾	3	3½	4	¾	4
2	4	⅝	2¾	3	8	⅝	3	3¼	8	⅝	4
2½	4	⅝	3	3¼	8	¾	3¼	3¾	8	¾	4½
3	4	⅝	3	3½	8	¾	3½	4	8	¾	4¾
3½	8	⅝	3	3½	8	¾	3¾	4¼	8	⅞	5¼
4	8	⅝	3	3½	8	¾	3¾	4¼	8	⅞	5¼
5	8	¾	3¼	3¾	8	¾	4	4½	8	⅞	5½
6	8	¾	3¼	3¾	12	¾	4¼	4¾	12	⅞	5¾
8	8	¾	3½	4	12	⅞	4¾	5¼	12	1	6½
10	12	⅞	3¾	4½	16	1	5¼	6	16	1⅛	7¼
12	12	⅞	4	4½	16	1⅛	5¾	6½	16	1¼	7¾
14	12	1	4¼	5	20	1⅛	6	6¾	20	1¼	8
16	16	1	4½	5¼	20	1¼	6½	7¼	20	1⅜	8½
18	16	1⅛	4¾	5¾	24	1¼	6¾	7½	24	1⅜	8¾
20	20	1⅛	5¼	6	24	1¼	7	8	24	1½	9½
24	20	1¼	5¾	6¾	24	1½	7¾	9	24	1¾	10½

LENGTH OF BOLTS FOR PIPE FLANGES

Nominal Pipe Size	600 LB. FLANGES			900 LB. FLANGES			1500 LB. FLANGES			2500 LB. FLANGES		
	Number of Bolts	Diam. of Bolts	Length of Stud Bolt	Number of Bolts	Diam. of Bolts	Length of Stud Bolts	Number of Bolts	Diam. of Bolts	Length of Stud Bolts	Number of Bolts	Diam. of Bolts	Length of Stud Bolts
½	4	½	3	4	¾	4	4	¾	4	4	¾	4¾
¾	4	⅝	3¼	4	¾	4¼	4	¾	4¼	4	¾	4¾
1	4	⅝	3½	4	⅞	4¾	4	⅞	4¾	4	⅞	5¼
1¼	4	⅝	3¾	4	⅞	4¾	4	⅞	4¾	4	1	5¾
1½	4	¾	4	4	1	5¼	4	1	5¼	4	1⅛	6½
2	8	⅝	4	8	⅞	5½	8	⅞	5½	8	1	6¾
2½	8	¾	4½	8	1	6	8	1	6	8	1⅛	7½
3	8	¾	4¾	8	⅞	5½	8	1⅛	6¾	8	1¼	8½.
3½	8	⅞	5¼	—	—	—	—	—	—	—	—	—
4	8	⅞	5½	8	1⅛	6½	8	1¼	7½	8	1½	9¾
5	8	1	6¼	8	1¼	7¼	8	1½	9½	8	1¾	11½
6	12	1	6½	12	1⅛	7½	12	1⅜	10	8	2	13½
8	12	1⅛	7½	12	1⅜	8½	12	1⅝	11¼	12	2	15
10	16	1¼	8¼	16	1⅜	9	12	1⅞	13¼	12	2½	19
12	20	1¼	8½	20	1⅜	9¾	16	2	14¾	12	2¾	21
14	20	1⅜	9	20	1½	10½	16	2¼	16	—	—	—
16	20	1½	9¾	20	1⅝	11	16	2½	17½	—	—	—
18	20	1⅝	10½	20	1⅞	12¾	16	2¾	19¼	—	—	—
20	24	1⅝	11¼	20	2	13½	16	3	21	—	—	—
24	24	1⅞	12¾	20	2½	17	16	3½	24	—	—	—

CAST STEEL FLANGED FITTINGS
CENTER TO FACE
OF
ELBOW, TEE, OR CROSS

150 LB.		300 LB.	
Size	Center to Face	Size	Center to Face
1 1/2	4	1 1/2	4 1/2
2	4 1/2	2	5
2 1/2	5	2 1/2	5 1/2
3	5 1/2	3	6
3 1/2	6	3 1/2	6 1/2
4	6 1/2	4	7
5	7 1/2	5	8
6	8	6	8 1/2
8	9	8	10
10	11	10	11 1/2
12	12	12	13
14	14	14	15
16	15	16	16 1/2
18	16 1/2	18	18
20	18	20	19 1/2
24	22	24	22 1/2

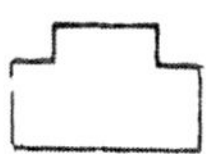

PIPE SIZE	CENTER to FACE LONG RADIUS	SHORT RAIDUS	CENTER to FACE LONG RAIDUS	PIPE SIZE	CENTER TO END
1/2	1 1/2		5/8	1/2	1
3/4	1 1/8		7/16	3/4	1 1/8
1	1 1/2	1	7/8	1	1 1/2
1 1/4	1 7/8	1 1/4	1	1 1/4	1 7/8
1 1/2	2 1/4	1 1/2	1 1/8	1 1/2	2 1/4
2	3	2	1 1/4	2	2 1/2
2 1/2	3 3/4	2 1/2	1 3/4	2 1/2	3
3	4 1/2	3	1 7/8	3	3 3/8
3 1/2	5 1/4	3 1/2	2 1/4	3 1/2	3 3/4
4	6	4	2 1/2	4	4 1/8
5	7 1/2	5	3 1/8	5	4 7/8
6	9	6	3 3/4	6	5 5/8
8	12	8	5	8	7
10	15	10	6 1/4	10	8 1/2
12	18	12	7 1/2	12	10
14	21	14	8 3/4	14	11
16	24	16	10	16	12
18	27	18	11 1/4	18	13 1/2
20	30	20	12 1/2	20	15
22	33		13 1/2	22	16 1/2
24	36		15	24	17
26	39		16	26	19 1/2
30	45		18 1/2	30	22
34	51		21	34	25
36	54		22 1/4	36	26 1/2
42	63		26	42	

TAKE UP OF WELDING REDUCERS

CONCENTRIC OR ECCENTRIC
STANDARD WEIGHT

PIPE SIZE	LENGTH	WEIGHT
¾x ⅜	1½	.21
¾x ½	1½	.21
1 x ⅜	2	.30
1 x ½	2	.31
1 x ¾	2	.31
1¼x ½	2	.43
1¼x ¾	2	.44
1¼x1	2	.45
1½x ½	2½	.45
1½x ¾	2½	.48
1½x1	2½	.53
1½x1¼	2½	.57
2 x ¾	3	.73
2 x1	3	.82
2 x1¼	3	.87
2 x1½	3	.90
2½x1	3½	1.30
2½x1¼	3½	1.47
2½x1½	3½	1.51
2½x2	3½	1.60
3 x1¼	3½	1.70
3 x1½	3½	1.89
3 x2	3½	2.00

TAKE UP OF WELDING REDUCERS

CONCENTRIC OR ECCENTRIC
STANDARD WEIGHT

PIPE SIZE	LENGTH	WEIGHT
3 x2½	3½	2.16
3½x1¼	4	2.35
3½x1½	4	2.52
3½x2	4	2.71
3½x2½	4	2.96
3½x3	4	3.05
4 x1½	4	2.73
4 x2	4	3.17
4 x2½	4	3.34
4 x3	4	3.50
4 x3½	4	3.61
5 x2	5	5.05
5 x2½	5	5.52
5 x3	5	5.73
5 x3½	5	5.86
5 x4	5	5.99
6 x2½	5½	7.61
6 x3	5½	8.00
6 x3½	5½	8.14
6 x4	5½	8.19
6 x5	5½	8.65
8 x3½	6	12.8
8 x4	6	13.1

CONCENTRIC OR ECCENTRIC
STANDARD WEIGHT

PIPE SIZE	LENGTH	WEIGHT
8x5	6	13.4
8x6	6	13.9
10x4	7	21.1
10x5	7	21.8
10x6	7	22.3
10x8	7	23.2
12x5	8	30.5
12x6	8	31.1
12x8	8	32.1
12x10	8	33.4
14x6	13	55.8
14x8	13	57.2
14x10	13	60.4
14x12	13	63.4
16x8	14	70.2
16x10	14	72.9
16x12	14	75.6
16x14	14	77.5
18x10	15	86.9
18x12	15	89.2
18x14	15	90.9
18x16	15	94.0
20x12	20	134
20x14	20	135
20x16	20	138
20x18	20	142
22x14	20	148
22x16	20	151
22x18	20	154

TAKE UP OF WELDING REDUCERS

CONCENTRIC OR ECCENTRIC
STANDARD WEIGHT

PIPE SIZE	LENGTH	WEIGHT	
		CONC.	ECC.
22x20	20	157	
24x16	20	160	
24x18	20	163	
24x20	20	167	
26x18	24	200	
26x20	24	200	
26x22	24	200	
26x24	24	200	
30x20	24	220	
30x24	24	220	
30x26	24	220	
30x28	24	220	
34x24	24	270	229
34x26	24	270	237
34x30	24	270	253
34x32	24	270	261
36x24	24	340	237
36x26	24	340	245
36x30	24	340	261
36x32	24	340	269
36x34	24	340	277
42x24	24	260	
42x26	24	270	
42x30	24	285	
42x32	24	295	
42x34	24	300	
42x36	24	310	

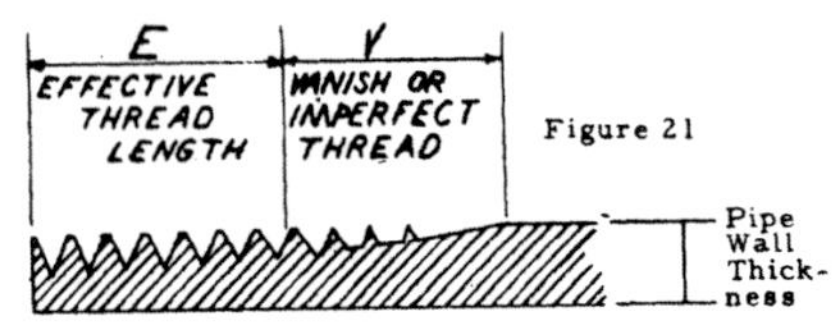

NOMINAL PIPE SIZE (INCHES)	NUMBER OF THREADS (PER INCH)	LENGTH OF THREADS (E + V) (INCHES)	EFFECTIVE LENGTH OF THREADS (E) (INCHES)
1/8	27	7/16	1/4
1/4	18	5/8	7/16
3/8	18	5/8	7/16
1/2	14	13/16	9/16
3/4	14	13/16	9/16
1	11 1/2	1	11/16
1 1/4	11 1/2	1	11/16
1 1/2	11 1/2	1 1/32	3/4
2	11 1/2	1 1/16	3/4
2 1/2	8	1 9/16	1 1/8
3	8	1 5/8	1 1/4
4	8	1 3/4	1 5/16
6	8	1 15/16	1 1/2
8	8	2 3/16	1 3/4
10	8	2 3/8	1 15/16
12	8	2 9/16	2 1/8

THREADED ELBOWS, TEES AND CROSSES
3000# FORGED STEEL

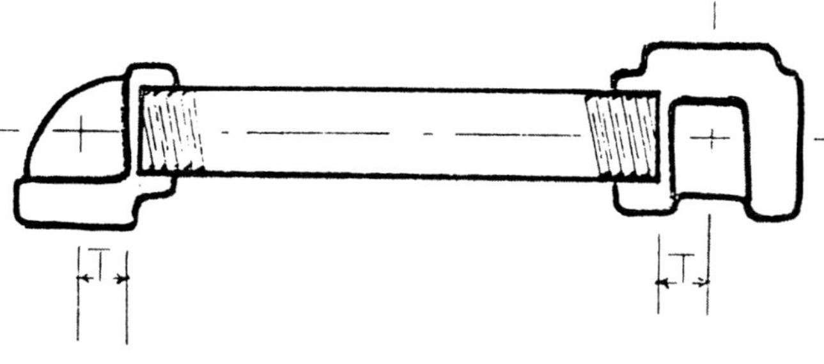

FITTING SIZE	CENTER TO END DIMENSION	TAKE UP
1/4"	1"	9/16"
3/8"	1 1/8"	11/16"
1/2"	1 5/16"	3/4"
3/4"	1 1/2"	15/16"
1"	1 3/4"	1 1/16"
1 1/2"	2 3/8"	1 5/8"
2"	2 1/2"	1 3/4"
2 1/2"	3 3/8"	2 1/4"
3"	3 3/4"	2 1/2"
4"	4 1/2"	2 11/16"

Take up is the distance from the center
of the fitting to the end of the pipe.

RING NUMBERS FOR RING JOINT FLANGES

150 Lb CLASS

PIPE SIZE	RING NO.	Number	STUD BOLTS Size	Length
1	R 15	4	1/2	3
1 1/4	R 17	4	1/2	3
1 1/2	R 19	4	1/2	3 1/4
2	R 22	4	5/8	3 1/2
2 1/2	R 25	4	5/8	3 3/4
3	R 29	4	5/8	4
4	R 36	8	5/8	4
5	R 40	8	3/4	4 1/4
6	R 43	8	3/4	4 1/4
8	R 48	8	3/4	4 1/2
10	R 52	12	7/8	5
12	R 56	12	7/8	5
14	R 59	12	1	5 1/2
16	R 64	16	1	5 3/4
18	R 68	16	1 1/8	6 1/4
20	R 72	20	1 1/8	6 1/2
24	R 76	20	1 1/4	7 1/4

300 Lb. CLASS

PIPE SIZE	RING NO.	Number	STUD BOLTS Size	Length
1/2	R 11	4	1/2	3
3/4	R 13	4	5/8	3 1/4
1	R 16	4	5/8	3 1/2
1 1/4	R 18	4	5/8	3 1/2
1 1/2	R 20	4	3/4	4
2	R 23	8	5/8	4
2 1/2	R 26	8	3/4	4 1/2
3	R 31	8	3/4	4 3/4
4	R 37	8	3/4	5
5	R 41	8	3/4	5 1/4
6	R 45	12	3/4	5 1/2
8	R 49	12	7/8	6
10	R 53	16	1	6 3/4
12	R 57	16	1 1/8	7 1/4
14	R 61	20	1 1/8	7 1/2
16	R 65	20	1 1/4	8
18	R 69	24	1 1/4	8 1/4
20	R 73	24	1 1/2	8 3/4
24	R 77	24	1 1/2	10

<u>RING NUMBERS FOR RING JOINT FLANGES</u>

<u>400 Lb. CLASS</u>

PIPE SIZE	RING NO.	Number	STUD BOLTS Size	Length
4	R 37	8	7/8	5 1/2
6	R 45	12	7/8	6
8	R 49	12	1	6 3/4
10	R 53	16	1 1/8	7 1/2
12	R 57	16	1 1/4	8
14	R 61	20	1 1/4	8 1/4
16	R 65	20	1 3/8	9
18	R 69	24	1 3/8	9
20	R 73	24	1 1/2	9 3/4
24	R 77	24	1 3/4	11

<u>600 Lb. CLASS</u>

PIPE SIZE	RING NO.	Number	STUD BOLTS Size	Length
1/2	R 11	4	1/2	3
3/4	R 13	4	5/8	3 1/4
1	R 16	4	5/8	3 1/2
1 1/4	R 18	4	5/8	3 3/4
1 1/2	R 20	4	3/4	4
2	R 23	8	5/8	4 1/4
2 1/2	R 26	8	3/4	4 3/4
3	R 31	8	3/4	5
4	R 37	8	7/8	3/4
5	R 41	8	1	6 1/2
6	R 45	12	1	6 3/4
8	R 49	12	1 1/8	7 3/4
10	R 53	16	1 1/4	8 1/2
12	R 57	20	1 1/4	8 3/4
14	R 61	20	1 3/8	9 1/4
16	R 65	20	1 1/2	10
18	R 69	20	1 5/8	10 3/4
20	R 73	24	1 5/8	11 1/2
24	R 77	24	1 7/8	13 1/4

RING NUMBERS FOR RING JOINT FLANGES

900 Lb. CLASS

PIPE SIZE	RING NO.	Number	STUD BOLTS Size	Length
3	R 31	8	7/8	5 3/4
4	R 37	8	1 1/8	6 3/4
5	R 41	8	1 1/4	7 1/2
6	R 45	12	1 1/8	7 1/2
8	R 49	12	1 3/8	8 3/4
10	R 53	16	1 3/8	9 1/4
12	R 57	20	1 3/8	10
14	R 62	20	1 1/2	11
16	R 66	20	1 5/8	11 1/2
18	R 70	20	1 7/8	13 1/4
20	R 74	20	2	14
24	R 78	20	2 1/2	17 3/4

1500 Lb. CLASS

PIPE SIZE	RING NO.	Number	STUD BOLTS Size	Length
1/2	R 12	4	3/4	4
3/4	R 14	4	3/4	4 1/4
1	R 16	4	7/8	4 3/4
1 1/4	R 18	4	7/8	4 3/4
1 1/2	R 20	4	1	5 1/4
2	R 24	8	7/8	5 3/4
2 1/2	R 27	8	1	6 1/4
3	R 35	8	1 1/8	7
4	R 39	8	1 1/4	7 3/4
5	R 44	8	1 1/2	9 3/4
6	R 46	12	1 3/8	10 1/4
8	R 50	12	1 5/8	11 3/4
10	R 54	12	1 7/8	13 1/2
12	R 58	16	2	15 1/4
14	R 63	16	2 1/4	16 3/4
16	R 67	16	2 1/2	18 1/2
18	R 71	16	2 3/4	20 1/4
20	R 75	16	3	22 1/4
24	R 79	16	3 1/2	25 1/2

<u>RING NUMBERS FOR RING JOINT FLANGES</u>

<u>2500 Lb. CLASS</u>

PIPE SIZE	RING NO.	Number	STUD BOLTS Size	Length
1/2	R 13	4	3/4	4 3/4
3/4	R 16	4	3/4	4 3/4
1	R 18	4	7/8	5 1/4
1 1/4	R 21	4	1	6
1 1/2	R 23	4	1 1/8	6 3/4
2	R 26	8	1	7
2 1/2	R 28	8	1 1/8	7 3/4
3	R 32	8	1 1/4	8 3/4
4	R 38	8	1 1/2	10 1/4
5	R 42	8	1 3/4	12 1/4
6	R 47	8	2	14
8	R 51	12	2	15 1/2
10	R 55	12	2 1/2	20
12	R 60	12	2 3/4	22

DIMENSIONS OF MECHANICAL JOINT FITTINGS

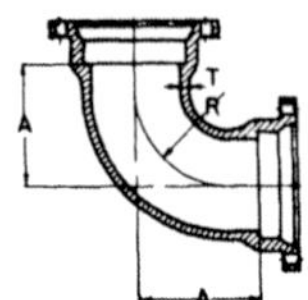 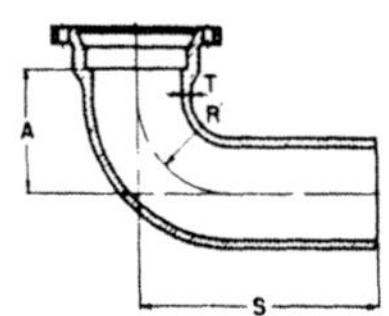 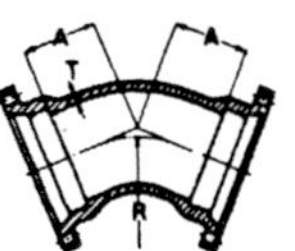 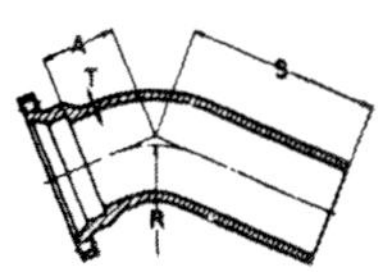

90° BENDS

Size	Class	Dimensions in Inches			
		T	R	A	S
3	250	.48	4.0	5.50	13.5
4	250	.52	4.5	6.50	14.5
6	250	.55	6.0	8.00	16.0
8	250	.60	7.0	9.00	17.0
10	250	.68	9.0	11.00	19.0
12	250	.75	10.0	12.00	20.0
14	150	.66	11.5	14.00	22.0
16	150	.70	12.5	15.00	23.0
18	150	.75	14.0	16.50	24.5
20	150	.80	15.5	18.00	26.0
24	150	.89	18.5	22.00	30.0

45° BENDS

Size	Class	Dimensions in Inches			
		T	R	A	S
3	250	.48	3.62	3.0	11.0
4	250	.52	4.81	4.0	12.0
6	250	.55	7.25	5.0	13.0
8	250	.60	8.44	5.5	13.5
10	250	.68	10.88	6.5	14.5
12	250	.75	13.25	7.5	15.5
14	150	.66	12.06	7.5	15.5
16	150	.70	13.25	8.0	16.0
18	150	.75	14.50	8.5	16.5
20	150	.80	16.88	9.5	17.5
24	150	.89	18.12	11.0	19.0

DIMENSIONS OF MECHANICAL JOINT FITTINGS

TEES AND CROSSES

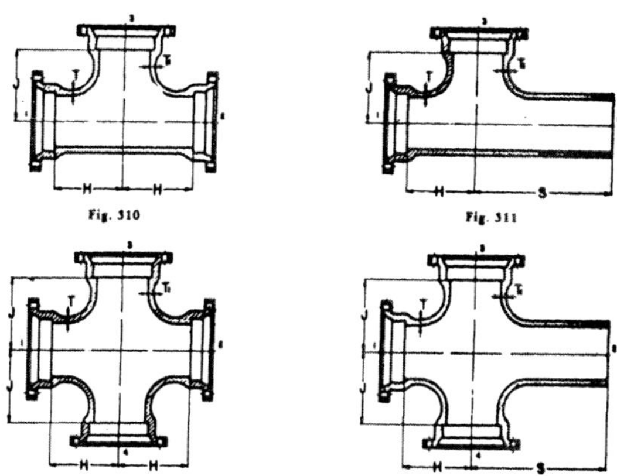

Fig. 310   Fig. 311

Size		Dimensions in Inches					
Run	Branch	Class	T	T₁	H	J	S
3	2	250	.48	.38	5.5	5.5	13.5
3	2½	250	.48	.38	5.5	5.5	13.5
3	3	250	.48	.48	5.5	5.5	13.5
4	2	250	.52	.38	6.5	6.5	14.5
4	2½	250	.52	.38	6.5	6.5	14.5
4	3	250	.52	.48	6.5	6.5	14.5
4	4	250	.52	.52	6.5	6.5	14.5
6	2	250	.55	.38	8	8	16
6	2½	250	.55	.38	8	8	16
6	3	250	.55	.48	8	8	16
6	4	250	.55	.52	8	8	16
6	6	250	.55	.55	8	8	16
8	3	250	.60	.48	9	9	17
8	4	250	.60	.52	9	9	17
8	6	250	.60	.55	9	9	17
8	8	250	.60	.60	9	9	17

DIMENSIONS OF MECHANICAL JOINT FITTINGS

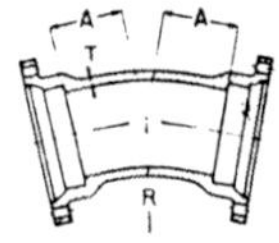 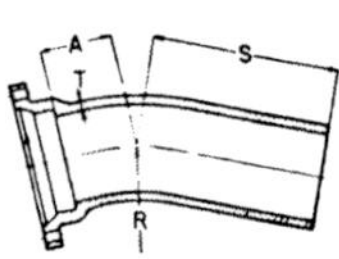 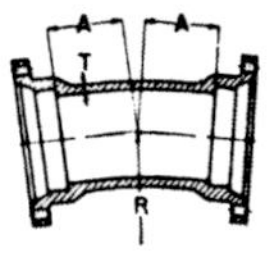 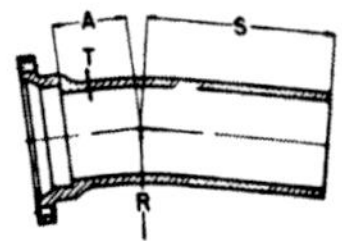

22 1/2° BENDS

Size	Class	Dimensions in Inches			
		T	R	A	S
3	250	.48	7.56	3.0	11.0
4	250	.52	10.06	4.0	12.0
6	250	.55	15.06	5.0	13.0
8	250	.60	17.62	5.5	13.5
10	250	.68	22.62	6.5	14.5
12	250	.75	27.62	7.5	15.5
14	150	.66	25.12	7.5	15.5
16	150	.70	27.62	8.0	16.0
18	150	.75	30.19	8.5	16.5
20	150	.80	35.19	9.5	17.5
24	150	.89	37.69	11.0	19.0

11 1/4° BENDS

Size	Class	Dimensions in Inches			
		T	R	A	S
3	250	.48	15.25	3.0	11.0
4	250	.52	20.31	4.0	12.0
6	250	.55	30.50	5.0	13.0
8	250	.60	35.50	5.5	13.5
10	250	.68	45.69	6.5	14.5
12	250	.75	55.81	7.5	15.5
14	150	.66	50.75	7.5	15.5
16	150	.70	55.81	8.0	16.0
18	150	.75	60.94	8.5	16.5
20	150	.80	71.06	9.5	17.5
24	150	.89	76.12	11.0	19.0

WRENCH AND BOLT SIZE EQUIVALENTS

BOLT SIZE	WRENCH SIZE
1/2"	7/8"
5/8"	1 1/16"
3/4"	1 1/4"
7/8"	1 7/16"
1"	1 5/8"
1 1/4"	2"
1 1/2"	2 3/8"
1 3/4"	2 3/4"
2"	3 1/8"

WRENCH SIZE = (1 1/2 X BOLT SIZE) + 1/8"

BOLT SIZE = (WRENCH SIZE - 1/8") X 2/3

PIPE SIZE	O.D.	CIR.	2 PARTS	4 PARTS	8 PARTS	16 PARTS
2"	2.375"	7 1/2"	3 3/4"	1 7/8"	15/16"	1/2"
2 1/2	2.875	9	4 1/2	2 1/4	1 1/8	9/16
3	3.5	11	5 1/2	2 3/4	1 3/8	11/16
4	4.5	14 1/8	7 1/16	3 1/2	1 3/4	7/8
5	5.56	17 1/2	8 3/4	4 3/8	2 3/16	1 1/8
6	6.625	20 7/8	10 7/16	5 3/16	2 5/8	1 5/16
8	8.625	27 1/8	13 9/16	6 3/4	3 3/8	1 11/16
10	10.75	33 3/4	16 7/8	8 7/16	4 1/4	2 1/8
12	12.75	40 1/16	20	10	5	2 1/2
14	14	44	22	11	5 1/2	2 3/4
16	16	50 1/4	25 1/8	12 9/16	6 1/4	3 1/8
18	18	56 9/16	28 1/4	14 1/8	7 1/16	3 1/2
20	20	62 13/16	31 7/16	15 11/16	7 7/8	3 15/16
22	22	69 1/8	34 9/16	17 1/4	8 5/8	4 5/16
24	24	75 13/16	37 11/16	18 7/8	9 7/16	4 3/4
30	30	94 1/4	47 1/8	23 9/16	11 3/4	5 7/8
36	36	113 1/8	56 9/16	28 1/4	14 1/8	7 1/16

<u>USEFUL FORMULAS</u>

1. Run of offset = cosecant of degree of turn times amount of offset.

2. Travel of offset = cotangent of degree of turn times amount of offset.

3. Amount of offset = Run pipe times sine of degree of turn or travel of offset times tangent of degree of turn.

4. Take up of welding elbow = tangent of one-half the degree of turn times radius of the elbow.

5. Angle of cut for mitered turn = one-half the degree of turn.

6. Cut back for miter = tangent of degree of cut times one-half the O.D. of the pipe. (Hacksaw type cut)

7. Cut back for miter = tangent of degree of cut times one-half the I.D. of the pipe. (Radial cut)

8. Center to center length between two 45 elbows of rolling offset = length of hypotenuse formed by set against roll multiplied by 1.414.

9. Center to center length of rolling offset (90) elbows = length of hypotenuse formed by set against roll.

10. Amount of pipe in a bend = .01745 times number of degrees in bend times radius of the bend.

11. Arc length on back of welding elbow = .01745 times number of degrees of turn times outside radius of elbow.

12. Arc length in throat of elbow = .01745 times number of degrees of turn times inside radius of elbow.

13. To change Fahreheit temperature to Centigrade: Subtract 32 from the Fahrenheit reading, multiply by 5, and divide the answer by 9.

14. To change Centigrade temperature to Fahrenheit: Multiply the Centigrade reading by 9, divide the answer by 5, then add 32 degrees.

15. To find the area of a rectangle, multiply the length of the rectangle times its width.

16. To find the area of a triangle, multiply the height of the triangle times its base and divide by two.

17. To find the area of a circle, multiply 3.1416 times the radius times the radius.

18. To find the volume of a tank, multiply the area of its base times the height.

ABBREVIATIONS USED IN THE PIPING TRADE

Am't. - amount
A. S. A. - American Standards
   Association
A. S. D. - American Standard
   Drilled
B. W. - Buttweld
Bev. - Bevel
Blk. - Blank
B. & S. - Bell and Spigot
B. C. - Bolt circle
C. I. - Cast Iron
C. S. - Cast Steel
C. to F. - Center to Face
C to C. - Center to center
C. L. - Centerline
Conc. - Concentric
Conn. - Connection
Cosec. - Cosecant
C. V. - Control Valve
C. S. O. - Car Seal Open
C. B. - Catch Basin
Cos. - Cosine
Circ. - Circumference
Cot. - Cotangent
Deg. - Degrees
Disch. - Discharge
Dwg. - Drawing
Det. - Detail
El. or Elev. - Elevation
Exh. Stm. - Exhaust Steam
Ecc. - Eccentric
Ex. Stg. - Extra Strong
Fld. - Field
Fnd. - Foundation
F. F. - Full Face or Flat Face
F. to F. - Face to Face

F. & D. - Faced and Drilled
Fig. - Figure
F. W. - Field weld
Flg. - Flange
Galv. - Galvanized
Ga. Va. - Gate Valve
G. G. - Gauge Glass
Gl. Va. - Globe Valve
H. P. - High Point or High
   Pressure
I. D. Inside Diameter
L. R. - Long Radius
Lubr. - Lubricate
Mach. - Machine
M. M. - Millimeter
Mk. - Mark
Max. - Maximum
Min. - Minimum
No. - Number
O. D. Outside Diameter
O. S. & Y. - Outside Stem and
   Yoke
P. V. - Pressure Vessel
P. D. - Pitch Diameter
P. S. I. - Pounds per Square
   Inch
Pc. - Piece
Pl. - Plate
Req'd. - Required
R. - Radius
R. F. - Raised Face
Red. - Reducer
R. T. J. Ring Type Joint
Rev. - Revision
Reg. - Regular
Sec. - Section

ABBREVIATIONS USED IN THE PIPING TRADE

Sin. - Sine

Sh. # - Sheet Number

S.O. Flg. - Slip on flange

Sch. - Schedule

S.R. - Short Radius

Ser. - Series

Scr'd. - Screwed

Sq. - Square

Std. Wt. - Standard weight

Tk. - Tank

T.L. - Threadolet

T.W. - Thermowell

T.I. - Temperature Indicator

T.C. - Thermocouple

Tan. or Tang. - Tangent

Temp. - Temperature or Temporary

W.O.G. - Water, Oil or Gas

W.N. Flg. - Welding Neck Flange

YD. No. - Yard Number

MULTIPLY	BY	TO OBTAIN
acres	43,560	square feet
acres	4840	square yards
B.T.U. per min.	.02356	horse power
bushels	1.244	cubic feet
bushels	2150	cubic inches
bushels	.03524	cubic meters
bushels	4	pecks
bushels	64	pints (dry)
bushels	32	quarts (dry)
centimeters	.3397	inches
centimeters	.01	meters
centimeters	393.7	mils
centimeters	10	millimeters
cubic feet	62.43	pounds of water
cubic feet	1728	cubic inches
cubic feet	.02832	cubic meters
cubic feet	.03704	cubic yards
cubic feet	7.481	gallons
cubic feet	28.32	liters
cubic feet	59.84	pints (Liq.)
cubic feet	29.92	quarts (Liq.)

<u>CONVERSION TABLES</u>

<u>MULTIPLY</u>	<u>BY</u>	<u>TO OBTAIN</u>
cubic feet per min.	.1247	gallons per sec.
cubic feet per min.	62.4	lbs. of water per mi
cubic inches	16.39	cubic centimeters
cubic inches	.03463	pints (Liq.)
cubic inches	.01732	quarts (Liq.)
cubic yards	27	cubic feet
cubic yards	46,656	cubic inches
cubic yards	.7646	cubic meters
cubic yards	202.0	gallons
cubic yards	764.6	liters
cubic yards	1616	pints (Liq.)
cubic yards	807.9	quarts (Liq.)
degrees (angle)	60	minutes
degrees (angle)	.01745	radians
degrees (angle)	3600	seconds
feet of water	.8826	inches of mercury
feet of water	62.43	pounds per sq. ft.
feet of water	.4335	pounds per sq. in.
gallons	8.345	pounds of water
gallons	3785	cubic centimeters
gallons	.1337	cubic feet

<u>CONVERSION TABLES</u>

<u>MULTIPLY</u>	<u>BY</u>	<u>TO OBTAIN</u>
gallons	231	cubic inches
gallons	3.785	cubic centimeters
gallons	8	pints (Liq.)
gallons	4	quarts (Liq.)
grams	.03527	ounces
inches of mercury	1.133	feet of water
inches of mercury	70.73	pounds per sq. ft.
inches of mercury	.4912	pounds per sq. in.
inches of water	.07355	inches of mercury
inches of water	5.204	pounds per sq. ft.
inches of water	.03613	pounds per sq. in.
kilometers	3281	feet
kilometers	.6214	miles
kilometers	1093.6	yards
meters	1.0936	yards
miles	5280	feet
miles	1.6093	kilometers
miles	1760	yards
miles per hour	88	feet per min.
miles per hour	1.467	feet per sec.
miles per hour	1.6093	kilometers per hr.

<u>CONVERSION TABLES</u>

<u>MULTIPLY</u>	<u>BY</u>	<u>TO OBTAIN</u>
miles per hour	.8684	knots per hr.
miles per hour	26.82	meters per min.
months	30.42	days
months	730	hours
months	43,800	minutes
ounces	8	drams
ounces	437.5	grains
ounces	28.35	grams
ounces	.0625	pounds
pints (Liq.)	28.87	cubic inches
pounds	7000	grains
pounds	453.6	grams
pounds	16	ounces
pounds of water	.01602	cubic feet
pounds of water	27.68	cubic inches
pounds of water	.1198	gallons
quarts	32	fluid ounces
quarts (Liq.)	57.75	cubic inches
rods	16.5	feet
square feet	144	square inches
square miles	640	acres
square yards	9	square feet

Conversion Table — English to Metric

(approximate to within 2%)

When You know	Multiply by	To Find	Symbol
inches	25	millimeters	mm
inches	2.5	centimeters	cm
feet	0.3	meters	m
yards	0.9	meters	m
fathoms	1.8	meters	m
miles	1.6	kilometers	km
square inches	6.5	square centimeters	cm²
square feet	0.092	square meters	m²
square yards	0.84	square meters	m²
acres	0.4	hectares	ha
square miles	2.6	square kilometers	km²
cubic inches	16.4	cubic centimeters	cm³
cubic feet	0.028	cubic meters	m³
cubic yards	0.76	cubic meters	m³
ounces (fluid)	30	milliliters	mL
pints ⎫ liquid	0.47	liters	L
quarts ⎭	0.95	liters	L
gallons	3.8	liters	L
pints ⎫ dry	0.55	liters	L
quarts ⎭	1.1	liters	L
bushels	35	liters	L
bushels	0.035	cubic meters	m³
avoirdupois ounces	28	grams	g
avoirdupois pounds	0.45	kilograms	kg
short tons (2000 lb)	0.91	metric tons	t
long tons (2240 lb)	1.02	metric tons	t

Fahrenheit temperature — 32 X 0.56 = Celsius temperature

Conversion Table — Metric to English

(approximate to within 2%)

When You Know	Multiply by	To Find
millimeters	0.04	inches
centimeters	0.4	inches
meters	3.3	feet
meters	1.1	yards
kilometers	0.62	miles
square centimeters	0.155	square inches
square meters	10.8	square feet
square meters	1.2	square yards
hectares	2.5	acres
square kilometers	0.39	square miles
cubic centimeters	0.06	cubic inches
cubic meters	35	cubic feet
cubic meters	1.3	cubic yards
milliliters	0.034	ounces (fluid)
liters	2.1	pints } liquid
liters	1.06	quarts } liquid
liters	0.26	gallons
liters	1.8	pints } dry
liters	0.9	quarts } dry
cubic meters	28	bushels
grams	0.035	avoirdupois ounce
kilograms	2.2	avoirdupois poun
metric tons	1.1	short tons (2000
metric tons	1.0	long tons (2240 l

Celsius temperature X 1.8 + 32 = Fahrenheit temperature

PIPING SYMBOLS
Flanges

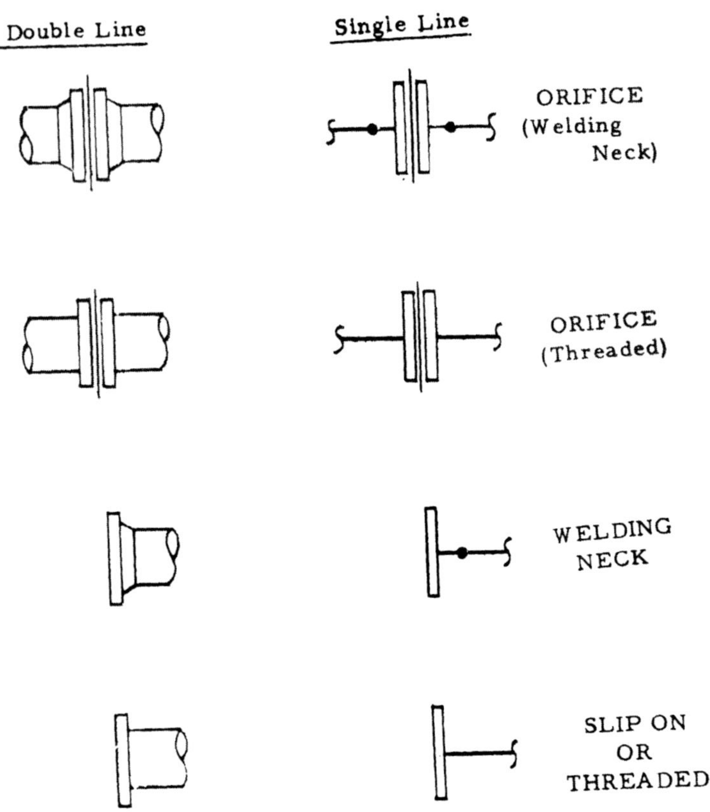

PIPING SYMBOLS

Valves

Double Line	Single Line	
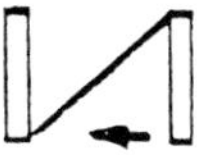	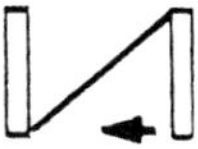	CHECK (Flanged)
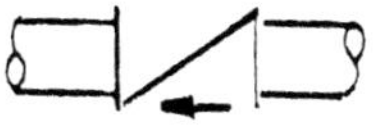	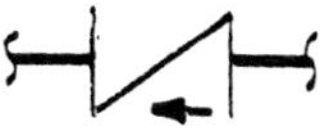	CHECK (Threaded)
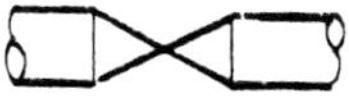	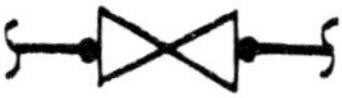	GATE (Welding)
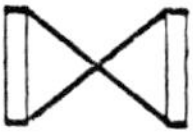	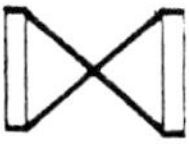	GATE (Flanged)
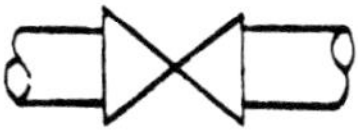	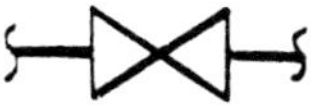	GATE (Threaded)

PIPING SYMBOLS

Valves

Double Line	Single Line	
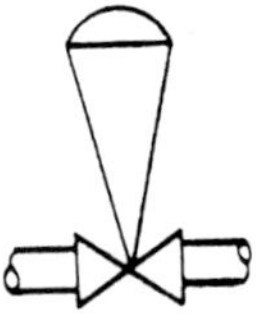	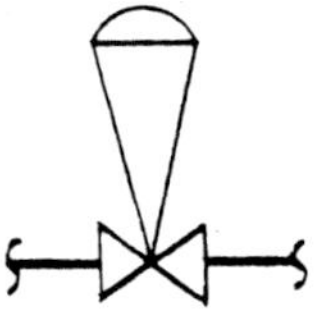	THREADED CONTROL VALVE
		FLANGED CONTROL VALVE
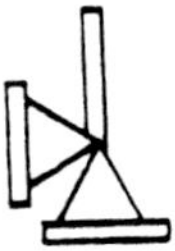	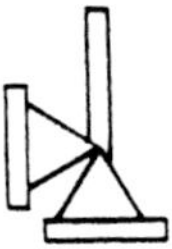	FLANGED SAFETY VALVE
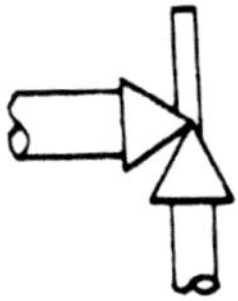	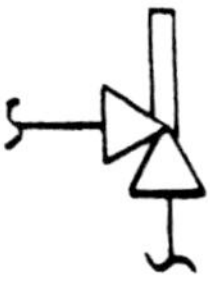	THREADED SAFETY VALVE

PIPING SYMBOLS
Tees

Single Line	Double Line	

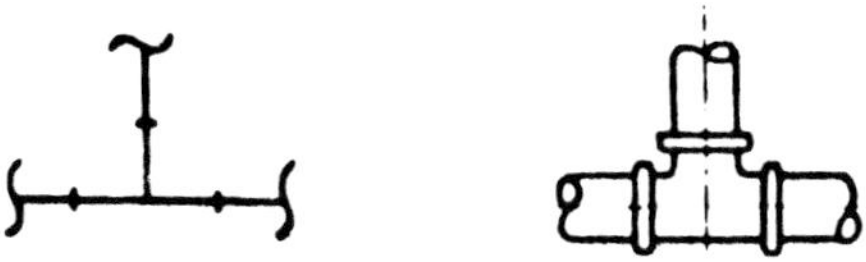

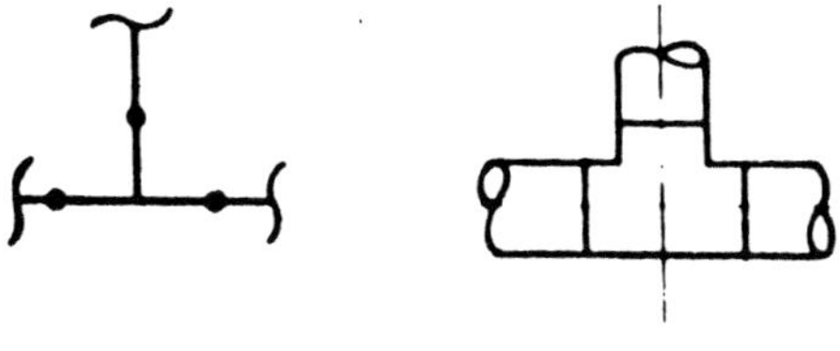

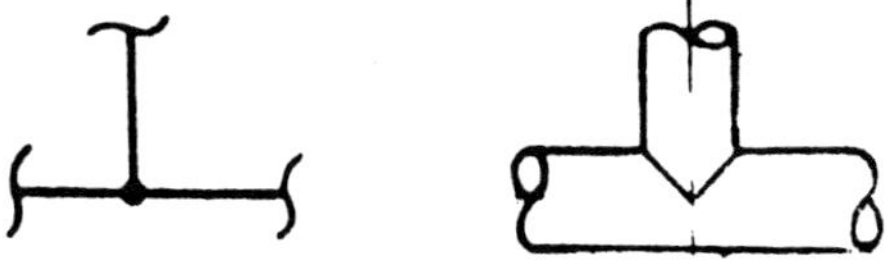

PIPING SYMBOLS
Elbows

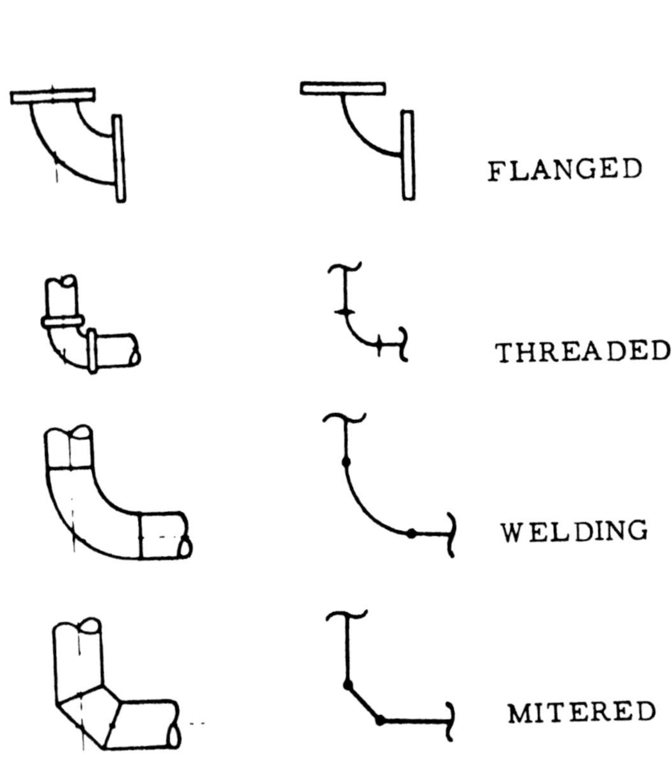

PIPING SYMBOLS
Swages

Double Line	Single Line	
		ECCENTRIC (Welding)
	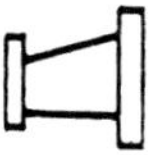	ECCENTRIC (Flanged)
	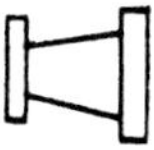	CONCENTRIC (Flanged)
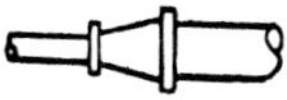		CONCENTRIC (Threaded)
		CONCENTRIC (Welding)

PIPING SYMBOLS
Miscellaneous

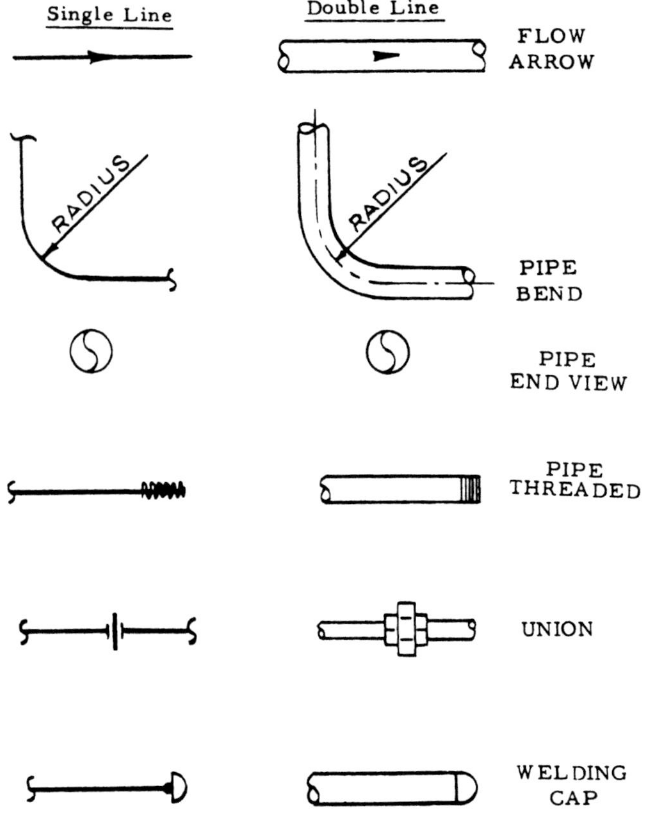

PIPING SYMBOLS
Bell and Spigot

<u>Double Line</u>   Single Line

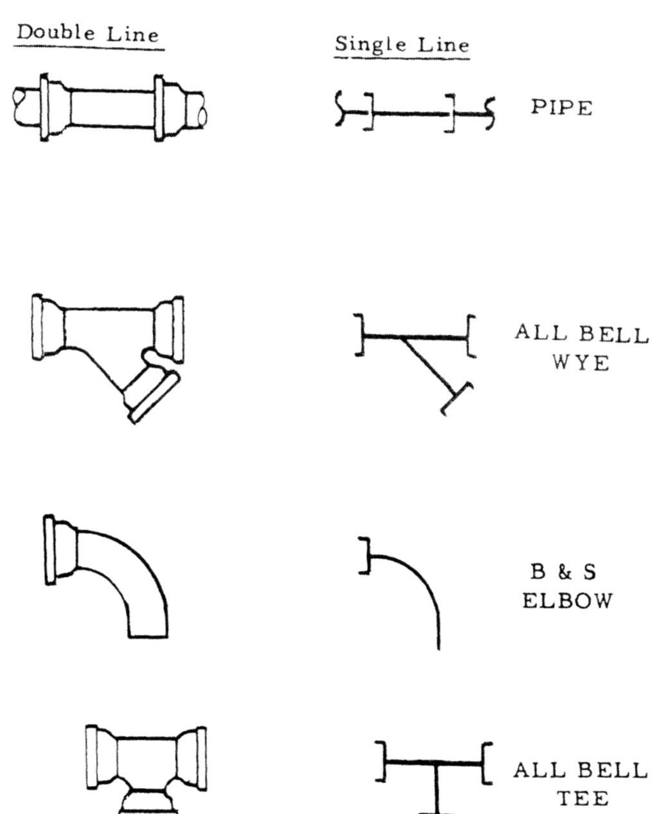

PIPING SYMBOLS
Bell and Spigot

Double Line	Single Line	
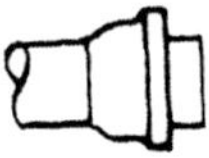	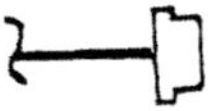	PLUG
		SPIGOT REDUCER
		ALL BEI L REDUCER
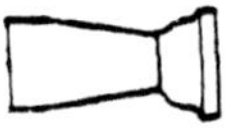		B & S REDUCER
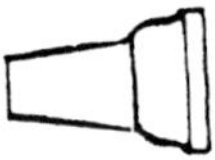		B & S INCREASER

TABLES OF

TRIGONOMETRIC

FUNCTIONS

0°

M	Sine	Cosine	Tan.	Cotan.	Secant	Cosec.	M
0	.00000	1.0000	.00000	Infinite	1.0000	Infinite	60
1	.00029	.0000	.00029	3437.7	.0000	3437.7	59
2	.00058	.0000	.00058	1718.9	.0000	1718.9	58
3	.00087	.0000	.00087	1145.9	.0000	1145.9	57
4	.00116	.0000	.00116	859.44	.0000	859.44	56
5	.00145	1.0000	.00145	687.55	1.0000	687.55	55
6	.00174	.0000	.00174	572.96	.0000	572.96	54
7	.00204	.0000	.00204	491.11	.0000	491.11	53
8	.00233	.0000	.00233	429.72	.0000	429.72	52
9	.00262	.0000	.00262	381.97	.0000	381.97	51
10	.00291	.99999	.00291	343.77	1.0000	343.77	50
11	.00320	.99999	.00320	312.52	.0000	312.52	49
12	.00349	.99999	.00349	286.48	.0000	286.48	48
13	.00378	.99999	.00378	264.44	.0000	264.44	47
14	.00407	.99999	.00407	245.55	.0000	245.55	46
15	.00436	.99999	.00436	229.18	1.0000	229.18	45
16	.00465	.99999	.00465	214.86	.0000	214.86	44
17	.00494	.99999	.00494	202.22	.0000	202.22	43
18	.00524	.99999	.00524	190.98	.0000	190.99	42
19	.00553	.99998	.00553	180.93	.0000	180.93	41
20	.00582	.99998	.00582	171.88	1.0000	171.89	40
21	.00611	.99998	.00611	163.70	.0000	163.70	39
22	.00640	.99998	.00640	156.26	.0000	156.26	38
23	.00669	.99998	.00669	149.46	.0000	149.47	37
24	.00698	.99997	.00698	143.24	.0000	143.24	36
25	.00727	.99997	.00727	137.51	1.0000	137.51	35
26	.00756	.99997	.00756	132.22	.0000	132.22	34
27	.00785	.99997	.00785	127.32	.0000	127.32	33
28	.00814	.99997	.00814	122.77	.0000	122.78	32
29	.00843	.99996	.00844	118.54	.0000	118.54	31
30	.00873	.99996	.00873	114.59	1.0000	114.59	30
31	.00902	.99996	.00902	110.89	.0000	110.90	29
32	.00931	.99996	.00931	107.43	.0000	107.43	28
33	.00960	.99995	.00960	104.17	.0000	104.17	27
34	.00989	.99995	.00989	101.11	.0000	101.11	26
35	.01018	.99995	.01018	98.218	1.0000	98.223	25
36	.01047	.99994	.01047	95.489	.0000	95.495	24
37	.01076	.99994	.01076	92.908	.0000	92.914	23
38	.01105	.99994	.01105	90.463	.0001	90.469	22
39	.01134	.99993	.01134	88.143	.0001	88.149	21
40	.01163	.99993	.01164	85.940	1.0001	85.946	20
41	.01193	.99993	.01193	83.843	.0001	83.849	19
42	.01222	.99992	.01222	81.847	.0001	81.853	18
43	.01251	.99992	.01251	79.943	.0001	79.950	17
44	.01280	.99992	.01280	78.126	.0001	78.133	16
45	.01309	.99991	.01309	76.390	1.0001	76.396	15
46	.01338	.99991	.01338	74.729	.0001	74.736	14
47	.01367	.99991	.01367	73.139	.0001	73.146	13
48	.01396	.99990	.01396	71.615	.0001	71.622	12
49	.01425	.99990	.01425	70.153	.0001	70.160	11
50	.01454	.99989	.01454	68.750	1.0001	68.757	10
51	.01483	.99989	.01484	67.402	.0001	67.409	9
52	.01512	.99988	.01513	66.105	.0001	66.113	8
53	.01542	.99988	.01542	64.858	.0001	64.866	7
54	.01571	.99988	.01571	63.657	.0001	63.664	6
55	.01600	.99987	.01600	62.499	1.0001	62.507	5
56	.01629	.99987	.01629	61.383	.0001	61.391	4
57	.01658	.99987	.01658	60.306	.0001	60.314	3
58	.01687	.99986	.01687	59.266	.0001	59.274	2
59	.01716	.99985	.01716	58.261	.0001	58.270	1
60	.01745	.99985	.01745	57.290	1.0001	57.299	0
M.	Cosine	Sine	Cotan.	Tan.	Cosec.	Secant	M

89°

1°

M	Sine	Cosine	Tan.	Cotan.	Secant	Cosec.	M
0	.01745	.99985	.01745	57.290	1.0001	57.299	60
1	.01774	.99984	.01775	56.350	.0001	56.359	59
2	.01803	.99984	.01804	55.441	.0001	55.450	58
3	.01832	.99983	.01833	54.561	.0002	54.570	57
4	.01861	.99983	.01862	53.708	.0002	53.718	56
5	.01891	.99982	.01891	52.882	1.0002	52.891	55
6	.01920	.99981	.01920	52.081	.0002	52.090	54
7	.01949	.99981	.01949	51.303	.0002	51.313	53
8	.01978	.99980	.01978	50.548	.0002	50.558	52
9	.02007	.99980	.02007	49.816	.0002	49.826	51
10	.02036	.99979	.02036	49.104	1.0002	49.114	50
11	.02065	.99979	.02066	48.412	.0002	48.422	49
12	.02094	.99978	.02095	47.739	.0002	47.750	48
13	.02123	.99977	.02124	47.085	.0002	47.096	47
14	.02152	.99977	.02153	46.449	.0002	46.460	46
15	.02181	.99976	.02182	45.829	1.0002	45.840	45
16	.02210	.99975	.02211	45.226	.0002	45.237	44
17	.02240	.99975	.02240	44.638	.0002	44.650	43
18	.02269	.99974	.02269	44.066	.0002	44.077	42
19	.02298	.99974	.02298	43.508	.0003	43.520	41
20	.02326	.99973	.02327	42.964	1.0003	42.976	40
21	.02356	.99972	.02357	42.433	.0003	42.445	39
22	.02385	.99971	.02386	41.916	.0003	41.928	38
23	.02414	.99971	.02415	41.410	.0003	41.423	37
24	.02443	.99970	.02444	40.917	.0003	40.930	36
25	.02472	.99969	.02473	40.436	1.0003	40.448	35
26	.02501	.99969	.02502	39.965	.0003	39.978	34
27	.02530	.99968	.02531	39.506	.0003	39.518	33
28	.02559	.99967	.02560	39.057	.0003	39.069	32
29	.02589	.99966	.02589	38.618	.0003	38.631	31
30	.02618	.99966	.02618	38.188	1.0003	38.201	30
31	.02647	.99965	.02648	37.769	.0003	37.782	29
32	.02676	.99964	.02677	37.358	.0003	37.371	28
33	.02705	.99963	.02706	36.956	.0004	36.969	27
34	.02734	.99963	.02735	36.563	.0004	36.576	26
35	.02763	.99962	.02764	36.177	1.0004	36.191	25
36	.02792	.99961	.02793	35.800	.0004	35.814	24
37	.02821	.99960	.02822	35.431	.0004	35.445	23
38	.02850	.99959	.02851	35.069	.0004	35.084	22
39	.02879	.99958	.02880	34.715	.0004	34.729	21
40	.02908	.99958	.02910	34.368	1.0004	34.382	20
41	.02937	.99957	.02939	34.027	.0004	34.042	19
42	.02967	.99956	.02968	33.693	.0004	33.708	18
43	.02996	.99955	.02997	33.366	.0004	33.381	17
44	.03025	.99954	.03026	33.045	.0004	33.060	16
45	.03054	.99953	.03055	32.730	1.0005	32.745	15
46	.03083	.99952	.03084	32.421	.0005	32.437	14
47	.03112	.99951	.03113	32.118	.0005	32.134	13
48	.03141	.99951	.03143	31.820	.0005	31.836	12
49	.03170	.99950	.03172	31.528	.0005	31.544	11
50	.03199	.99949	.03201	31.241	1.0005	31.257	10
51	.03228	.99948	.03230	30.960	.0005	30.976	9
52	.03257	.99947	.03259	30.683	.0005	30.699	8
53	.03286	.99946	.03288	30.411	.0005	30.428	7
54	.03315	.99945	.03317	30.145	.0005	30.161	6
55	.03344	.99944	.03346	29.882	1.0005	29.899	5
56	.03374	.99943	.03375	29.624	.0006	29.641	4
57	.03403	.99942	.03405	29.371	.0006	29.388	3
58	.03432	.99941	.03434	29.122	.0006	29.139	2
59	.03461	.99940	.03463	28.877	.0006	28.894	1
60	.03490	.99939	.03492	28.636	1.0006	28.654	0
M	Cosine	Sine	Cotan.	Tan.	Cosec.	Secant	M

88°

2°

M	Sine	Cosine	Tan.	Cotan.	Secant	Cosec.	M
0	.03490	.99939	.03492	28.636	1.0006	28.654	60
1	.03519	.99938	.03521	28.399	.0006	28.417	59
2	.03548	.99937	.03550	28.166	.0006	28.184	58
3	.03577	.99936	.03579	27.937	.0006	27.955	57
4	.03606	.99935	.03608	27.712	.0006	27.730	56
5	.03635	.99934	.03638	27.490	1.0007	27.508	55
6	.03664	.99933	.03667	27.271	.0007	27.290	54
7	.03693	.99932	.03696	27.056	.0007	27.075	53
8	.03722	.99931	.03725	26.845	.0007	26.864	52
9	.03751	.99930	.03754	26.637	.0007	26.655	51
10	.03781	.99928	.03783	26.432	1.0007	26.450	50
11	.03810	.99927	.03812	26.230	.0007	26.249	49
12	.03839	.99926	.03842	26.031	.0007	26.050	48
13	.03868	.99925	.03871	25.835	.0007	25.854	47
14	.03897	.99924	.03900	25.642	.0008	25.661	46
15	.03926	.99923	.03929	25.452	1.0008	25.471	45
16	.03955	.99922	.03958	25.264	.0008	25.284	44
17	.03984	.99921	.03987	25.080	.0008	25.100	43
18	.04013	.99919	.04016	24.898	.0008	24.918	42
19	.04042	.99918	.04045	24.718	.0008	24.739	41
20	.04071	.99917	.04075	24.542	1.0008	24.562	40
21	.04100	.99916	.04104	24.367	.0008	24.388	39
22	.04129	.99915	.04133	24.196	.0008	24.216	38
23	.04158	.99913	.04162	24.026	.0009	24.047	37
24	.04187	.99912	.04191	23.859	.0009	23.880	36
25	.04217	.99911	.04220	23.694	1.0009	23.716	35
26	.04246	.99910	.04249	23.532	.0009	23.553	34
27	.04275	.99908	.04279	23.372	.0009	23.393	33
28	.04304	.99907	.04308	23.214	.0009	23.235	32
29	.04333	.99906	.04337	23.058	.0009	23.079	31
30	.04362	.99905	.04366	22.904	1.0009	22.925	30
31	.04391	.99903	.04395	22.752	.0010	22.774	29
32	.04420	.99902	.04424	22.602	.0010	22.624	28
33	.04449	.99901	.04453	22.454	.0010	22.476	27
34	.04478	.99900	.04483	22.308	.0010	22.330	26
35	.04507	.99898	.04512	22.164	1.0010	22.186	25
36	.04536	.99897	.04541	22.022	.0010	22.044	24
37	.04565	.99896	.04570	21.881	.0010	21.904	23
38	.04594	.99894	.04599	21.742	.0010	21.765	22
39	.04623	.99893	.04628	21.606	.0011	21.629	21
40	.04652	.99892	.04657	21.470	1.0011	21.494	20
41	.04681	.99890	.04687	21.337	.0011	21.360	19
42	.04711	.99889	.04716	21.205	.0011	21.228	18
43	.04740	.99888	.04745	21.075	.0011	21.098	17
44	.04769	.99886	.04774	20.946	.0011	20.970	16
45	.04798	.99885	.04803	20.819	1.0011	20.843	15
46	.04827	.99883	.04832	20.693	.0012	20.717	14
47	.04856	.99882	.04862	20.569	.0012	20.593	13
48	.04885	.99881	.04891	20.446	.0012	20.471	12
49	.04914	.99879	.04920	20.325	.0012	20.350	11
50	.04943	.99878	.04949	20.205	1.0012	20.230	10
51	.04972	.99876	.04978	20.087	.0012	20.112	9
52	.05001	.99875	.05007	19.970	.0012	19.995	8
53	.05030	.99873	.05037	19.854	.0013	19.880	7
54	.05059	.99872	.05066	19.740	.0013	19.766	6
55	.05088	.99870	.05095	19.627	1.0013	19.653	5
56	.05117	.99869	.05124	19.515	.0013	19.541	4
57	.05146	.99867	.05153	19.405	.0013	19.431	3
58	.05175	.99866	.05182	19.296	.0013	19.322	2
59	.05204	.99864	.05212	19.188	.0013	19.214	1
60	.05234	.99863	.05241	19.081	1.0014	19.107	0
M	Cosine	Sine	Cotan.	Tan.	Cosec.	Secant	M

87°

3°

M	Sine	Cosine	Tan.	Cotan.	Secant	Cosec.	M
0	.05234	.99863	.05241	19.081	1.0014	19.107	60
1	.05263	.99861	.05270	18.975	.0014	19.002	59
2	.05292	.99860	.05299	18.871	.0014	18.897	58
3	.05321	.99858	.05328	18.768	.0014	18.794	57
4	.05350	.99857	.05357	18.665	.0014	18.692	56
5	.05379	.99855	.05387	18.564	1.0014	18.591	55
6	.05408	.99854	.05416	18.464	.0015	18.491	54
7	.05437	.99852	.05445	18.365	.0015	18.393	53
8	.05466	.99850	.05474	18.268	.0015	18.295	52
9	.05495	.99849	.05503	18.171	.0015	18.198	51
10	.05524	.99847	.05532	18.075	1.0015	18.103	50
11	.05553	.99846	.05562	17.980	.0015	18.008	49
12	.05582	.99844	.05591	17.886	.0016	17.914	48
13	.05611	.99842	.05620	17.793	.0016	17.821	47
14	.05640	.99841	.05649	17.701	.0016	17.730	46
15	.05669	.99839	.05678	17.610	1.0016	17.639	45
16	.05698	.99837	.05707	17.520	.0016	17.549	44
17	.05727	.99836	.05737	17.431	.0016	17.460	43
18	.05756	.99834	.05766	17.343	.0017	17.372	42
19	.05785	.99832	.05795	17.256	.0017	17.285	41
20	.05814	.99831	.05824	17.169	1.0017	17.198	40
21	.05843	.99829	.05853	17.084	.0017	17.113	39
22	.05872	.99827	.05883	16.999	.0017	17.028	38
23	.05902	.99826	.05912	16.915	.0017	16.944	37
24	.05931	.99824	.05941	16.832	.0018	16.861	36
25	.05960	.99822	.05970	16.750	1.0018	16.779	35
26	.05989	.99820	.05999	16.668	.0018	16.698	34
27	.06018	.99819	.06029	16.587	.0018	16.617	33
28	.06047	.99817	.06058	16.507	.0018	16.538	32
29	.06076	.99815	.06087	16.428	.0018	16.459	31
30	.06105	.99813	.06116	16.350	1.0019	16.380	30
31	.06134	.99812	.06145	16.272	.0019	16.303	29
32	.06163	.99810	.06175	16.195	.0019	16.226	28
33	.06192	.99808	.06204	16.119	.0019	16.150	27
34	.06221	.99806	.06233	16.043	.0019	16.075	26
35	.06250	.99804	.06262	15.969	1.0019	16.000	25
36	.06279	.99803	.06291	15.894	.0020	15.926	24
37	.06308	.99801	.06321	15.821	.0020	15.853	23
38	.06337	.99799	.06350	15.748	.0020	15.780	22
39	.06366	.99797	.06379	15.676	.0020	15.708	21
40	.06395	.99795	.06408	15.605	1.0020	15.637	20
41	.06424	.99793	.06437	15.534	.0021	15.566	19
42	.06453	.99791	.06467	15.464	.0021	15.496	18
43	.06482	.99790	.06496	15.394	.0021	15.427	17
44	.06511	.99788	.06525	15.325	.0021	15.358	16
45	.06540	.99786	.06554	15.257	1.0021	15.290	15
46	.06569	.99784	.06583	15.189	.0022	15.222	14
47	.06598	.99782	.06613	15.122	.0022	15.155	13
48	.06627	.99780	.06642	15.056	.0022	15.089	12
49	.06656	.99778	.06671	14.990	.0022	15.023	11
50	.06685	.99776	.06700	14.924	1.0022	14.958	10
51	.06714	.99774	.06730	14.860	.0023	14.893	9
52	.06743	.99772	.06759	14.795	.0023	14.829	8
53	.06772	.99770	.06788	14.732	.0023	14.765	7
54	.06801	.99768	.06817	14.668	.0023	14.702	6
55	.06830	.99766	.06846	14.606	1.0023	14.640	5
56	.06859	.99764	.06876	14.544	.0024	14.578	4
57	.06888	.99762	.06905	14.482	.0024	14.517	3
58	.06918	.99760	.06934	14.421	.0024	14.456	2
59	.06947	.99758	.06963	14.361	.0024	14.395	1
60	.06976	.99756	.06993	14.301	1.0024	14.335	0

M	Cosine	Sine	Cotan.	Tan.	Cosec.	Secant	M

86°

4°

M	Sine	Cosine	Tan.	Cotan.	Secant	Cosec.	M
0	.06976	.99756	.06993	14.301	1.0024	14.335	60
1	.07005	.99754	.07022	14.241	.0025	14.276	59
2	.07034	.99752	.07051	14.182	.0025	14.217	58
3	.07063	.99750	.07080	14.123	.0025	14.159	57
4	.07092	.99748	.07110	14.065	.0025	14.101	56
5	.07121	.99746	.07139	14.008	1.0025	14.043	55
6	.07150	.99744	.07168	13.951	.0026	13.986	54
7	.07179	.99742	.07197	13.894	.0026	13.930	53
8	.07208	.99740	.07226	13.838	.0026	13.874	52
9	.07237	.99738	.07256	13.782	.0026	13.818	51
10	.07266	.99736	.07285	13.727	1.0026	13.763	50
11	.07295	.99733	.07314	13.672	.0027	13.708	49
12	.07324	.99731	.07343	13.617	.0027	13.654	48
13	.07353	.99729	.07373	13.553	.0027	13.600	47
14	.07382	.99727	.07402	13.510	.0027	13.547	46
15	.07411	.99725	.07431	13.457	1.0027	13.494	45
16	.07440	.99723	.07460	13.404	.0028	13.441	44
17	.07469	.99721	.07490	13.351	.0028	13.389	43
18	.07498	.99718	.07519	13.299	.0028	13.337	42
19	.07527	.99716	.07548	13.248	.0028	13.286	41
20	.07556	.99714	.07577	13.197	1.0029	13.235	40
21	.07585	.99712	.07607	13.146	.0029	13.184	39
22	.07614	.99710	.07636	13.096	.0029	13.134	38
23	.07643	.99707	.07665	13.046	.0029	13.084	37
24	.07672	.99705	.07694	12.996	.0029	13.034	36
25	.07701	.99703	.07724	12.947	1.0030	12.985	35
26	.07730	.99701	.07753	12.898	1.0030	12.937	34
27	.07759	.99698	.07782	12.849	.0030	12.888	33
28	.07788	.99696	.07812	12.801	.0030	12.840	32
29	.07817	.99694	.07841	12.754	.0031	12.793	31
30	.07846	.99692	.07870	12.706	1.0031	12.745	30
31	.07875	.99689	.07899	12.659	.0031	12.698	29
32	.07904	.99687	.07929	12.612	.0031	12.652	28
33	.07933	.99685	.07958	12.566	.0032	12.606	27
34	.07962	.99682	.07987	12.520	.0032	12.560	26
35	.07991	.99680	.08016	12.474	1.0032	12.514	25
36	.08020	.99678	.08046	12.429	.0032	12.469	24
37	.08049	.99675	.08075	12.384	.0032	12.424	23
38	.08078	.99673	.08104	12.339	.0033	12.379	22
39	.08107	.99671	.08134	12.295	.0033	12.335	21
40	.08136	.99668	.08163	12.250	1.0033	12.291	20
41	.08165	.99666	.08192	12.207	.0033	12.248	19
42	.08194	.99664	.08221	12.163	.0034	12.204	18
43	.08223	.99661	.08251	12.120	.0034	12.161	17
44	.08252	.99659	.08280	12.077	.0034	12.118	16
45	.08281	.99656	.08309	12.035	1.0034	12.076	15
46	.08310	.99654	.08339	11.992	.0035	12.034	14
47	.08339	.99652	.08368	11.950	.0035	11.992	13
48	.08368	.99649	.08397	11.909	.0035	11.950	12
49	.08397	.99647	.08426	11.867	.0035	11.909	11
50	.08426	.99644	.08456	11.826	1.0036	11.868	10
51	.08455	.99642	.08485	11.785	.0036	11.828	9
52	.08484	.99639	.08514	11.745	.0036	11.787	8
53	.08513	.99637	.08544	11.704	.0036	11.747	7
54	.08542	.99634	.08573	11.664	.0037	11.707	6
55	.08571	.99632	.08602	11.625	1.0037	11.668	5
56	.08600	.99629	.08632	11.585	.0037	11.628	4
57	.08629	.99627	.08661	11.546	1.0037	11.589	3
58	.08658	.99624	.08690	11.507	.0038	11.550	2
59	.08687	.99622	.08719	11.468	.0038	11.512	1
60	.08715	.99619	.08749	11.430	1.0038	11.474	0

M	Cosine	Sine	Cotan.	Tan.	Cosec.	Secant	M

85°

5°

M	Sine	Cosine	Tan.	Cotan.	Secant	Cosec.	M
0	.08715	.99619	.08749	11.430	1.0038	11.474	60
1	.08744	.99617	.08778	11.392	.0038	11.436	59
2	.08773	.99614	.08807	11.354	.0039	11.398	58
3	.08802	.99612	.08837	11.316	.0039	11.360	57
4	.08831	.99609	.08866	11.279	.0039	11.323	56
5	.08860	.99607	.08895	11.242	1.0039	11.286	55
6	.08889	.99604	.08925	11.205	.0040	11.249	54
7	.08918	.99601	.08954	11.168	.0040	11.213	53
8	.08947	.99599	.08983	11.132	.0040	11.176	52
9	.08976	.99596	.09013	11.095	.0040	11.140	51
10	.09005	.99594	.09042	11.059	1.0041	11.104	50
11	.09034	.99591	.09071	11.024	.0041	11.069	49
12	.09063	.99588	.09101	10.988	.0041	11.033	48
13	.09092	.99586	.09130	10.953	.0041	10.998	47
14	.09121	.99583	.09159	10.918	.0042	10.963	46
15	.09150	.99580	.09189	10.883	1.0042	10.929	45
16	.09179	.99578	.09218	10.848	.0042	10.894	44
17	.09208	.99575	.09247	10.814	.0043	10.860	43
18	.09237	.99572	.09277	10.780	.0043	10.826	42
19	.09266	.99570	.09306	10.746	.0043	10.792	41
20	.09295	.99567	.09335	10.712	1.0043	10.758	40
21	.09324	.99564	.09365	10.678	.0044	10.725	39
22	.09353	.99562	.09394	10.645	.0044	10.692	38
23	.09382	.99559	.09423	10.612	.0044	10.659	37
24	.09411	.99556	.09453	10.579	.0044	10.626	36
25	.09440	.99553	.09482	10.546	1.0045	10.593	35
26	.09469	.99551	.09511	10.514	.0045	10.561	34
27	.09498	.99548	.09541	10.481	.0045	10.529	33
28	.09527	.99545	.09570	10.449	.0046	10.497	32
29	.09556	.99542	.09599	10.417	.0046	10.465	31
30	.09584	.99540	.09629	10.385	1.0046	10.433	30
31	.09613	.99537	.09658	10.354	.0046	10.402	29
32	.09642	.99534	.09688	10.322	.0047	10.371	28
33	.09671	.99531	.09717	10.291	.0047	10.340	27
34	.09700	.99528	.09746	10.260	.0047	10.309	26
35	.09729	.99525	.09776	10.229	1.0048	10.278	25
36	.09758	.99523	.09805	10.199	.0048	10.248	24
37	.09787	.99520	.09834	10.168	.0048	10.217	23
38	.09816	.99517	.09864	10.138	.0048	10.187	22
39	.09845	.99514	.09893	10.108	.0049	10.157	21
40	.09874	.99511	.09922	10.078	1.0049	10.127	20
41	.09903	.99508	.09952	10.048	.0049	10.098	19
42	.09932	.99505	.09981	10.019	.0050	10.068	18
43	.09961	.99503	.10011	9.9893	.0050	10.039	17
44	.09990	.99500	.10040	9.9601	.0050	10.010	16
45	.10019	.99497	.10069	9.9310	1.0050	9.9812	15
46	.10048	.99494	.10099	9.9021	.0051	9.9525	14
47	.10077	.99491	.10128	9.8734	.0051	9.9239	13
48	.10106	.99488	.10158	9.8448	.0051	9.8955	12
49	.10134	.99485	.10187	9.8164	.0052	9.8672	11
50	.10163	.99482	.10216	9.7882	1.0052	9.8391	10
51	.10192	.99479	.10246	9.7601	.0052	9.8112	9
52	.10221	.99476	.10275	9.7322	.0053	9.7834	8
53	.10250	.99473	.10305	9.7044	.0053	9.7558	7
54	.10279	.99470	.10334	9.6768	.0053	9.7283	6
55	.10308	.99467	.10363	9.6493	1.0053	9.7010	5
56	.10337	.99464	.10393	9.6220	.0054	9.6739	4
57	.10366	.99461	.10422	9.5949	.0054	9.6469	3
58	.10395	.99458	.10452	9.5679	.0054	9.6200	2
59	.10424	.99455	.10481	9.5411	.0055	9.5933	1
60	.10453	.99452	.10510	9.5144	1.0055	9.5668	0
M	Cosine	Sine	Cotan.	Tan.	Cosec.	Secant	M

84°

6°

M	Sine	Cosine	Tan.	Cotan.	Secant	Cosec.	M
0	.10453	.99452	.10510	9.5144	1.0055	9.5668	60
1	.10482	.99449	.10540	.4878	.0055	.5404	59
2	.10511	.99446	.10569	.4614	.0056	.5141	58
3	.10540	.99443	.10599	.4351	.0056	.4880	57
4	.10568	.99440	.10628	.4090	.0056	.4620	56
5	.10597	.99437	.10657	9.3831	1.0057	9.4362	55
6	.10626	.99434	.10687	.3572	.0057	.4105	54
7	.10655	.99431	.10716	.3315	.0057	.3850	53
8	.10684	.99428	.10746	.3060	.0057	.3596	52
9	.10713	.99424	.10775	.2806	.0058	.3343	51
10	.10742	.99421	.10805	9.2553	1.0058	9.3092	50
11	.10771	.99418	.10834	.2302	.0058	.2842	49
12	.10800	.99415	.10863	.2051	.0059	.2593	48
13	.10829	.99412	.10893	.1803	.0059	.2346	47
14	.10858	.99409	.10922	.1555	.0059	.2100	46
15	.10887	.99406	.10952	9.1309	1.0060	9.1855	45
16	.10916	.99402	.10981	.1064	.0060	.1612	44
17	.10944	.99399	.11011	.0821	.0060	.1370	43
18	.10973	.99396	.11040	.0579	.0061	.1129	42
19	.11002	.99393	.11069	.0338	.0061	.0890	41
20	.11031	.99390	.11099	9.0098	1.0061	9.0651	40
21	.11060	.99386	.11128	8.9860	.0062	.0414	39
22	.11089	.99383	.11158	.9623	.0062	.0179	38
23	.11118	.99380	.11187	.9387	.0062	8.9944	37
24	.11147	.99377	.11217	.9152	.0063	.9711	36
25	.11176	.99373	.11246	8.8918	1.0063	8.9479	35
26	.11205	.99370	.11276	.8686	.0063	.9248	34
27	.11234	.99367	.11305	.8455	.0064	.9018	33
28	.11262	.99364	.11335	.8225	.0064	.8790	32
29	.11291	.99360	.11364	.7996	.0064	.8563	31
30	.11320	.99357	.11393	8.7769	1.0065	8.8337	30
31	.11349	.99354	.11423	.7542	.0065	.8112	29
32	.11378	.99350	.11452	.7317	.0065	.7888	28
33	.11407	.99347	.11482	.7093	.0066	.7665	27
34	.11436	.99344	.11511	.6870	.0066	.7444	26
35	.11465	.99341	.11541	8.6648	1.0066	8.7223	25
36	.11494	.99337	.11570	.6427	.0067	.7004	24
37	.11523	.99334	.11600	.6208	.0067	.6786	23
38	.11551	.99330	.11629	.5989	.0067	.6569	22
39	.11580	.99327	.11659	.5772	.0068	.6353	21
40	.11609	.99324	.11688	8.5555	1.0068	8.6138	20
41	.11638	.99320	.11718	.5340	.0068	.5924	19
42	.11667	.99317	.11747	.5126	.0069	.5711	18
43	.11696	.99314	.11777	.4913	.0069	.5499	17
44	.11725	.99310	.11806	.4701	.0069	.5289	16
45	.11754	.99307	.11836	8.4489	1.0070	8.5079	15
46	.11783	.99303	.11865	.4279	.0070	.4871	14
47	.11811	.99300	.11895	.4070	.0070	.4663	13
48	.11840	.99296	.11924	.3862	.0071	.4457	12
49	.11869	.99293	.11954	.3655	.0071	.4251	11
50	.11898	.99290	.11983	8.3449	1.0071	8.4046	10
51	.11927	.99286	.12013	.3244	.0072	.3843	9
52	.11956	.99283	.12042	.3040	.0072	.3640	8
53	.11985	.99279	.12072	.2837	.0073	.3439	7
54	.12014	.99276	.12101	.2635	.0073	.3238	6
55	.12042	.99272	.12131	8.2434	1.0073	8.3039	5
56	.12071	.99269	.12160	.2234	.0074	.2840	4
57	.12100	.99265	.12190	.2035	.0074	.2642	3
58	.12129	.99262	.12219	.1837	.0074	.2446	2
59	.12158	.99258	.12249	.1640	.0075	.2250	1
60	.12187	.99255	.12278	8.1443	1.0075	8.2055	0
M	Cosine	Sine	Cotan.	Tan.	Cosec.	Secant	M

83°

M	Sine	Cosine	Tan.	Cotan.	Secant	Cosec.	M
0	.12187	.99255	.12278	8.1443	1.0075	8.2055	60
1	.12216	.99251	.12308	.1248	.0075	.1861	59
2	.12245	.99247	.12337	.1053	.0076	.1668	58
3	.12273	.99244	.12367	.0860	.0076	.1476	57
4	.12302	.99240	.12396	.0667	.0076	.1285	56
5	.12331	.99237	.12426	8.0476	1.0077	8.1094	55
6	.12360	.99233	.12456	.0285	.0077	.0905	54
7	.12389	.99229	.12485	.0095	.0078	.0717	53
8	.12418	.99226	.12515	7.9906	.0078	.0529	52
9	.12447	.99222	.12544	.9717	.0078	.0342	51
10	.12476	.99219	.12574	7.9530	1.0079	8.0156	50
11	.12504	.99215	.12603	.9344	.0079	7.9971	49
12	.12533	.99211	.12633	.9158	.0079	.9787	48
13	.12562	.99208	.12662	.8973	.0080	.9604	47
14	.12591	.99204	.12692	.8789	.0080	.9421	46
15	.12620	.99200	.12722	7.8606	1.0080	7.9240	45
16	.12649	.99197	.12751	.8424	.0081	.9059	44
17	.12678	.99193	.12781	.8243	.0081	.8879	43
18	.12706	.99189	.12810	.8062	.0082	.8700	42
19	.12735	.99186	.12840	.7882	.0082	.8522	41
20	.12764	.99182	.12869	7.7703	1.0082	7.8344	40
21	.12793	.99178	.12899	.7525	.0083	.8168	39
22	.12822	.99174	.12928	.7348	.0083	.7992	38
23	.12851	.99171	.12958	.7171	.0084	.7817	37
24	.12879	.99167	.12988	.6996	.0084	.7642	36
25	.12908	.99163	.13017	7.6821	1.0084	7.7469	35
26	.12937	.99160	.13047	.6646	.0085	.7296	34
27	.12966	.99156	.13076	.6473	.0085	.7124	33
28	.12995	.99152	.13106	.6300	.0085	.6953	32
29	.13024	.99148	.13136	.6129	.0086	.6783	31
30	.13053	.99144	.13165	7.5957	1.0086	7.6613	30
31	.13081	.99141	.13195	.5787	.0087	.6444	29
32	.13110	.99137	.13224	.5617	.0087	.6276	28
33	.13139	.99133	.13254	.5449	.0087	.6108	27
34	.13168	.99129	.13284	.5280	.0088	.5942	26
35	.13197	.99125	.13313	7.5113	1.0088	7.5776	25
36	.13226	.99121	.13343	.4946	.0089	.5611	24
37	.13254	.99118	.13372	.4780	.0089	.5446	23
38	.13283	.99114	.13402	.4615	.0089	.5282	22
39	.13312	.99110	.13432	.4451	.0090	.5119	21
40	.13341	.99106	.13461	7.4287	1.0090	7.4957	20
41	.13370	.99102	.13491	.4124	.0090	.4795	19
42	.13399	.99098	.13520	.3961	.0091	.4634	18
43	.13427	.99094	.13550	.3800	.0091	.4474	17
44	.13456	.99090	.13580	.3639	.0092	.4315	16
45	.13485	.99086	.13609	7.3479	1.0092	7.4156	15
46	.13514	.99083	.13639	.3319	.0092	.3998	14
47	.13543	.99079	.13669	.3160	.0093	.3840	13
48	.13571	.99075	.13698	.3002	.0093	.3683	12
49	.13600	.99071	.13728	.2844	.0094	.3527	11
50	.13629	.99067	.13757	7.2687	1.0094	7.3372	10
51	.13658	.99063	.13787	.2531	.0094	.3217	9
52	.13687	.99059	.13817	.2375	.0095	.3063	8
53	.13716	.99055	.13846	.2220	.0095	.2909	7
54	.13744	.99051	.13876	.2066	.0096	.2757	6
55	.13773	.99047	.13906	7.1912	1.0096	7.2604	5
56	.13802	.99043	.13935	.1759	.0097	.2453	4
57	.13831	.99039	.13965	.1607	.0097	.2302	3
58	.13860	.99035	.13995	.1455	.0097	.2152	2
59	.13888	.99031	.14024	.1304	.0098	.2002	1
60	.13917	.99027	.14054	7.1154	1.0098	7.1853	0
M	Cosine	Sine	Cotan.	Tan.	Cosec.	Secant	M

M	Sine	Cosine	Tan.	Cotan.	Secant	Cosec.	M
0	.13917	.99027	.14054	7.1154	1.0098	7.1853	60
1	.13946	.99023	.14084	.1004	.0099	.1704	59
2	.13975	.99019	.14113	.0854	.0099	.1557	58
3	.14004	.99015	.14143	.0706	.0099	.1409	57
4	.14032	.99010	.14173	.0558	.0100	.1263	56
5	.14061	.99006	.14202	7.0410	1.0100	7.1117	55
6	.14090	.99002	.14232	.0264	.0101	.0972	54
7	.14119	.98998	.14262	.0117	.0101	.0827	53
8	.14148	.98994	.14291	6.9972	.0102	.0683	52
9	.14176	.98990	.14321	.9827	.0102	.0539	51
10	.14205	.98986	.14351	6.9682	1.0102	7.0396	50
11	.14234	.98982	.14380	.9538	.0103	.0254	49
12	.14263	.98978	.14410	.9395	.0103	.0112	48
13	.14292	.98973	.14440	.9252	.0104	6.9971	47
14	.14320	.98969	.14470	.9110	.0104	.9830	46
15	.14349	.98965	.14499	6.8969	1.0104	6.9690	45
16	.14378	.98961	.14529	.8828	.0105	.9550	44
17	.14407	.98957	.14559	.8687	.0105	.9411	43
18	.14436	.98952	.14588	.8547	.0106	.9273	42
19	.14464	.98948	.14618	.8408	.0106	.9135	41
20	.14493	.98944	.14648	6.8269	1.0107	6.8998	40
21	.14522	.98940	.14677	.8131	.0107	.8861	39
22	.14551	.98936	.14707	.7993	.0107	.8725	38
23	.14579	.98931	.14737	.7856	.0108	.8589	37
24	.14608	.98927	.14767	.7720	.0108	.8454	36
25	.14637	.98923	.14796	6.7584	1.0109	6.8320	35
26	.14666	.98919	.14826	.7448	.0109	.8185	34
27	.14695	.98914	.14856	.7313	.0110	.8052	33
28	.14723	.98910	.14886	.7179	.0110	.7919	32
29	.14752	.98906	.14915	.7045	.0111	.7787	31
30	.14781	.98901	.14945	6.6911	1.0111	6.7655	30
31	.14810	.98897	.14975	.6779	.0111	.7523	29
32	.14838	.98893	.15004	.6646	.0112	.7392	28
33	.14867	.98889	.15034	.6514	.0112	.7262	27
34	.14896	.98884	.15064	.6383	.0113	.7132	26
35	.14925	.98880	.15094	6.6252	1.0113	6.7003	25
36	.14953	.98876	.15123	.6122	.0114	.6874	24
37	.14982	.98871	.15153	.5992	.0114	.6745	23
38	.15011	.98867	.15183	.5863	.0115	.6617	22
39	.15040	.98862	.15213	.5734	.0115	.6490	21
40	.15068	.98858	.15243	6.5605	1.0115	6.6363	20
41	.15097	.98854	.15272	.5478	.0116	.6237	19
42	.15126	.98849	.15302	.5350	.0116	.6111	18
43	.15155	.98845	.15332	.5223	.0117	.5985	17
44	.15183	.98840	.15362	.5097	.0117	.5860	16
45	.15212	.98836	.15391	6.4971	1.0118	6.5736	15
46	.15241	.98832	.15421	.4845	.0118	.5612	14
47	.15270	.98827	.15451	.4720	.0119	.5488	13
48	.15298	.98823	.15481	.4596	.0119	.5365	12
49	.15328	.98818	.15511	.4472	.0119	.5243	11
50	.15356	.98814	.15540	6.4348	1.0120	6.5121	10
51	.15385	.98809	.15570	.4225	.0120	.4999	9
52	.15413	.98805	.15600	.4103	.0121	.4878	8
53	.15442	.98800	.15630	.3980	.0121	.4757	7
54	.15471	.98796	.15659	.3859	.0122	.4637	6
55	.15500	.98791	.15689	6.3737	1.0122	6.4517	5
56	.15528	.98787	.15719	.3616	.0123	.4398	4
57	.15557	.98782	.15749	.3496	.0123	.4279	3
58	.15586	.98778	.15779	.3376	.0124	.4160	2
59	.15615	.98773	.15809	.3257	.0124	.4042	1
60	.15643	.98769	.15838	6.3137	1.0125	6.3924	0
M	Cosine	Sine	Cotan.	Tan.	Cosec.	Secant	M

M	Sine	Cosine	Tan.	Cotan.	Secant	Cosec.	M
0	.15643	.98769	.15838	6.3137	1.0125	6.3924	60
1	.15672	.98764	.15868	.3019	.0125	.3807	59
2	.15701	.98760	.15898	.2901	.0125	.3690	58
3	.15730	.98755	.15928	.2783	.0126	.3574	57
4	.15758	.98750	.15958	.2665	.0126	.3458	56
5	.15787	.98746	.15987	6.2548	1.0127	6.3343	55
6	.15816	.98741	.16017	.2432	.0127	.3228	54
7	.15844	.98737	.16047	.2316	.0128	.3113	53
8	.15873	.98732	.16077	.2200	.0128	.2999	52
9	.15902	.98727	.16107	.2085	.0129	.2885	51
10	.15931	.98723	.16137	6.1970	1.0129	6.2772	50
11	.15959	.98718	.16167	.1856	.0130	.2659	49
12	.15988	.98714	.16196	.1742	.0130	.2546	48
13	.16017	.98709	.16226	.1628	.0131	.2434	47
14	.16045	.98704	.16256	.1515	.0131	.2322	46
15	.16074	.98700	.16286	6.1402	1.0132	6.2211	45
16	.16103	.98695	.16316	.1290	.0132	.2100	44
17	.16132	.98690	.16346	.1178	.0133	.1990	43
18	.16160	.98685	.16376	.1066	.0133	.1880	42
19	.16189	.98681	.16405	.0955	.0134	.1770	41
20	.16218	.98676	.16435	6.0844	1.0134	6.1661	40
21	.16246	.98671	.16465	.0734	.0135	.1552	39
22	.16275	.98667	.16495	.0624	.0135	.1443	38
23	.16304	.98662	.16525	.0514	.0136	.1335	37
24	.16333	.98657	.16555	.0405	.0136	.1227	36
25	.16361	.98652	.16585	6.0296	1.0136	6.1120	35
26	.16390	.98648	.16615	.0188	.0137	.1013	34
27	.16419	.98643	.16644	.0080	.0137	.0906	33
28	.16447	.98638	.16674	5.9972	.0138	.0800	32
29	.16476	.98633	.16704	.9865	.0138	.0694	31
30	.16505	.98628	.16734	5.9758	1.0139	6.0588	30
31	.16533	.98624	.16764	.9651	.0139	.0483	29
32	.16562	.98619	.16794	.9545	.0140	.0379	28
33	.16591	.98614	.16824	.9439	.0140	.0274	27
34	.16619	.98609	.16854	.9333	.0141	.0170	26
35	.16648	.98604	.16884	5.9228	1.0141	6.0066	25
36	.16677	.98600	.16914	.9123	.0142	5.9963	24
37	.16705	.98595	.16944	.9019	.0142	.9860	23
38	.16734	.98590	.16973	.8915	.0143	.9758	22
39	.16763	.98585	.17003	.8811	.0143	.9655	21
40	.16791	.98580	.17033	5.8708	1.0144	5.9554	20
41	.16820	.98575	.17063	.8605	.0144	.9452	19
42	.16849	.98570	.17093	.8502	.0145	.9351	18
43	.16878	.98565	.17123	.8400	.0145	.9250	17
44	.16906	.98560	.17153	.8298	.0146	.9150	16
45	.16935	.98556	.17183	5.8196	1.0146	5.9049	15
46	.16964	.98551	.17213	.8095	.0147	.8950	14
47	.16992	.98546	.17243	.7994	.0147	.8850	13
48	.17021	.98541	.17273	.7894	.0148	.8751	12
49	.17050	.98536	.17303	.7794	.0148	.8652	11
50	.17078	.98531	.17333	5.7694	1.0149	5.8554	10
51	.17107	.98526	.17363	.7594	.0150	.8456	9
52	.17136	.98521	.17393	.7495	.0150	.8358	8
53	.17164	.98516	.17423	.7396	.0151	.8261	7
54	.17193	.98511	.17453	.7297	.0151	.8163	6
55	.17221	.98506	.17483	5.7199	1.0152	5.8067	5
56	.17250	.98501	.17513	.7101	.0152	.7970	4
57	.17279	.98496	.17543	.7004	.0153	.7874	3
58	.17307	.98491	.17573	.6906	.0153	.7778	2
59	.17336	.98486	.17603	.6809	.0154	.7683	1
60	.17365	.98481	.17633	5.6713	1.0154	5.7588	0
M	Cosine	Sine	Cotan.	Tan.	Cosec.	Secant	M

M	Sine	Cosine	Tan.	Cotan.	Secant	Cosec.	M
0	.17365	.98481	.17633	5.6713	1.0154	5.7588	60
1	.17393	.98476	.17663	.6616	.0155	.7493	59
2	.17422	.98471	.17693	.6520	.0155	.7398	58
3	.17451	.98465	.17723	.6425	.0156	.7304	57
4	.17479	.98460	.17753	.6329	.0156	.7210	56
5	.17508	.98455	.17783	5.6234	1.0157	5.7117	55
6	.17537	.98450	.17813	.6140	.0157	.7023	54
7	.17565	.98445	.17843	.6045	.0158	.6930	53
8	.17594	.98440	.17873	.5951	.0158	.6838	52
9	.17622	.98435	.17903	.5857	.0159	.6745	51
10	.17651	.98430	.17933	5.5764	1.0159	5.6653	50
11	.17680	.98425	.17963	.5670	.0160	.6561	49
12	.17708	.98419	.17993	.5578	.0160	.6470	48
13	.17737	.98414	.18023	.5485	.0161	.6379	47
14	.17766	.98409	.18053	.5393	.0162	.6288	46
15	.17794	.98404	.18083	5.5301	1.0162	5.6197	45
16	.17823	.98399	.18113	.5209	.0163	.6107	44
17	.17852	.98394	.18143	.5117	.0163	.6017	43
18	.17880	.98388	.18173	.5026	.0164	.5928	42
19	.17909	.98383	.18203	.4936	.0164	.5838	41
20	.17937	.98378	.18233	5.4845	1.0165	5.5749	40
21	.17966	.98373	.18263	.4755	.0165	.5660	39
22	.17995	.98368	.18293	.4665	.0166	.5572	38
23	.18023	.98362	.18323	.4575	.0166	.5484	37
24	.18052	.98357	.18353	.4486	.0167	.5396	36
25	.18080	.98352	.18383	5.4396	1.0167	5.5308	35
26	.18109	.98347	.18413	.4308	.0168	.5221	34
27	.18138	.98341	.18444	.4219	.0169	.5134	33
28	.18166	.98336	.18474	.4131	.0169	.5047	32
29	.18195	.98331	.18504	.4043	.0170	.4960	31
30	.18223	.98325	.18534	5.3955	1.0170	5.4874	30
31	.18252	.98320	.18564	.3868	.0171	.4788	29
32	.18281	.98315	.18594	.3780	.0171	.4702	28
33	.18309	.98309	.18624	.3694	.0172	.4617	27
34	.18338	.98304	.18654	.3607	.0172	.4532	26
35	.18366	.98299	.18684	5.3521	1.0173	5.4447	25
36	.18395	.98293	.18714	.3434	.0174	.4362	24
37	.18424	.98288	.18745	.3349	.0174	.4278	23
38	.18452	.98283	.18775	.3263	.0175	.4194	22
39	.18481	.98277	.18805	.3178	.0175	.4110	21
40	.18509	.98272	.18835	5.3093	1.0176	5.4026	20
41	.18538	.98267	.18865	.3008	.0176	.3943	19
42	.18507	.98261	.18895	.2923	.0177	.3860	18
43	.18595	.98256	.18925	.2839	.0177	.3777	17
44	.18624	.98250	.18955	.2755	.0178	.3695	16
45	.18652	.98245	.18985	5.2671	1.0179	5.3612	15
46	.18681	.98240	.19016	.2588	.0179	.3530	14
47	.18709	.98234	.19046	.2505	.0180	.3449	13
48	.18738	.98229	.19076	.2422	.0180	.3367	12
49	.18767	.98223	.19106	.2339	.0181	.3286	11
50	.18795	.98218	.19136	5.2257	1.0181	5.3205	10
51	.18824	.98212	.19166	.2174	.0182	.3124	9
52	.18852	.98207	.19197	.2092	.0182	.3044	8
53	.18881	.98201	.19227	.2011	.0183	.2963	7
54	.18909	.98196	.19257	.1929	.0184	.2883	6
55	.18938	.98190	.19287	5.1848	1.0184	5.2803	5
56	.18967	.98185	.19317	.1767	.0185	.2724	4
57	.18995	.98179	.19347	.1686	.0185	.2645	3
58	.19024	.98174	.19378	.1606	.0186	.2566	2
59	.19052	.98168	.19408	.1525	.0186	.2487	1
60	.19081	.98163	.19438	5.1445	1.0187	5.2408	0
M	Cosine	Sine	Cotan.	Tan.	Cosec.	Secant	M

11°

M	Sine	Cosine	Tan.	Cotan.	Secant	Cosec.	M
0	.19081	.98163	.19438	5.1445	1.0187	5.2408	60
1	.19109	.98157	.19468	.1366	.0188	.2330	59
2	.19138	.98152	.19498	.1286	.0188	.2252	58
3	.19166	.98146	.19529	.1207	.0189	.2174	57
4	.19195	.98140	.19559	.1128	.0189	.2097	56
5	.19224	.98135	.19589	5.1049	1.0190	5.2019	55
6	.19252	.98129	.19619	.0970	.0191	.1942	54
7	.19281	.98124	.19649	.0892	.0191	.1865	53
8	.19309	.98118	.19680	.0814	.0192	.1788	52
9	.19338	.98112	.19710	.0736	.0192	.1712	51
10	.19366	.98107	.19740	5.0658	1.0193	5.1636	50
11	.19395	.98101	.19770	.0581	.0193	.1560	49
12	.19423	.98095	.19800	.0504	.0194	.1484	48
13	.19452	.98090	.19831	.0427	.0195	.1409	47
14	.19480	.98084	.19861	.0350	.0195	.1333	46
15	.19509	.98078	.19891	5.0273	1.0196	5.1258	45
16	.19537	.98073	.19921	.0197	.0196	.1183	44
17	.19566	.98067	.19952	.0121	.0197	.1109	43
18	.19595	.98061	.19982	.0045	.0198	.1034	42
19	.19623	.98056	.20012	4.9969	.0198	.0960	41
20	.19652	.98050	.20042	4.9894	1.0199	5.0886	40
21	.19680	.98044	.20073	.9819	.0199	.0812	39
22	.19709	.98039	.20103	.9744	.0200	.0739	38
23	.19737	.98033	.20133	.9669	.0201	.0666	37
24	.19766	.98027	.20163	.9594	.0201	.0593	36
25	.19794	.98021	.20194	4.9520	1.0202	5.0520	35
26	.19823	.98016	.20224	.9446	.0202	.0447	34
27	.19851	.98010	.20254	.9372	.0203	.0375	33
28	.19880	.98004	.20285	.9298	.0204	.0302	32
29	.19908	.97998	.20315	.9225	.0204	.0230	31
30	.19937	.97992	.20345	4.9151	1.0205	5.0158	30
31	.19965	.97987	.20375	.9078	.0205	.0087	29
32	.19994	.97981	.20406	.9006	.0206	.0015	28
33	.20022	.97975	.20436	.8933	.0207	4.9944	27
34	.20051	.97969	.20466	.8860	.0207	.9873	26
35	.20079	.97963	.20497	4.8788	1.0208	4.9802	25
36	.20108	.97957	.20527	.8716	.0208	.9732	24
37	.20136	.97952	.20557	.8644	.0209	.9661	23
38	.20165	.97946	.20588	.8573	.0210	.9591	22
39	.20193	.97940	.20618	.8501	.0210	.9521	21
40	.20222	.97934	.20648	4.8430	1.0211	4.9452	20
41	.20250	.97928	.20679	.8359	.0211	.9382	19
42	.20279	.97922	.20709	.8288	.0212	.9313	18
43	.20307	.97916	.20739	.8217	.0213	.9243	17
44	.20336	.97910	.20770	.8147	.0213	.9175	16
45	.20364	.97904	.20800	4.8077	1.0214	4.9106	15
46	.20393	.97899	.20830	.8007	.0215	.9037	14
47	.20421	.97893	.20861	.7937	.0215	.8969	13
48	.20450	.97887	.20891	.7867	.0216	.8901	12
49	.20478	.97881	.20921	.7798	.0216	.8833	11
50	.20506	.97875	.20952	4.7728	1.0217	4.8765	10
51	.20535	.97869	.20982	.7659	.0218	.8697	9
52	.20563	.97863	.21012	.7591	.0218	.8630	8
53	.20592	.97857	.21043	.7522	.0219	.8563	7
54	.20620	.97851	.21073	.7453	.0220	.8496	6
55	.20649	.97845	.21104	4.7385	1.0220	4.8429	5
56	.20677	.97839	.21134	.7317	.0221	.8362	4
57	.20706	.97833	.21164	.7249	.0221	.8296	3
58	.20734	.97827	.21195	.7181	.0222	.8229	2
59	.20763	.97821	.21225	.7114	.0223	.8163	1
60	.20791	.97815	.21256	4.7046	1.0223	4.8097	0
M	Cosine	Sine	Cotan.	Tan.	Cosec.	Secant	M

78°

12°

M	Sine	Cosine	Tan.	Cotan.	Secant	Cosec.	M
0	.20791	.97815	.21256	4.7046	1.0223	4.8097	60
1	.20820	.97809	.21286	.6979	.0224	.8032	59
2	.20848	.97803	.21316	.6912	.0225	.7968	58
3	.20876	.97797	.21347	.6845	.0225	.7901	57
4	.20905	.97790	.21377	.6778	.0226	.7835	56
5	.20933	.97784	.21408	4.6712	1.0226	4.7770	55
6	.20962	.97778	.21438	.6646	.0227	.7706	54
7	.20990	.97772	.21468	.6580	.0228	.7641	53
8	.21019	.97766	.21499	.6514	.0228	.7576	52
9	.21047	.97760	.21529	.6448	.0229	.7512	51
10	.21076	.97754	.21560	4.6382	1.0230	4.7448	50
11	.21104	.97748	.21590	.6317	.0230	.7384	49
12	.21132	.97741	.21621	.6252	.0231	.7320	48
13	.21161	.97735	.21651	.6187	.0232	.7257	47
14	.21189	.97729	.21682	.6122	.0232	.7193	46
15	.21218	.97723	.21712	4.6057	1.0233	4.7130	45
16	.21246	.97717	.21742	.5993	.0234	.7067	44
17	.21275	.97711	.21773	.5928	.0234	.7004	43
18	.21303	.97704	.21803	.5864	.0235	.6942	42
19	.21331	.97698	.21834	.5800	.0235	.6879	41
20	.21360	.97692	.21864	4.5736	1.0236	4.6817	40
21	.21388	.97686	.21895	.5673	.0237	.6754	39
22	.21417	.97680	.21925	.5609	.0237	.6692	38
23	.21445	.97673	.21956	.5546	.0238	.6631	37
24	.21473	.97667	.21986	.5483	.0239	.6569	36
25	.21502	.97661	.22017	4.5420	1.0239	4.6507	35
26	.21530	.97655	.22047	.5357	.0240	.6446	34
27	.21559	.97648	.22078	.5294	.0241	.6385	33
28	.21587	.97642	.22108	.5232	.0241	.6324	32
29	.21615	.97636	.22139	.5169	.0242	.6263	31
30	.21644	.97630	.22169	4.5107	1.0243	4.6201	30
31	.21672	.97623	.22200	.5045	.0243	.6142	29
32	.21701	.97617	.22230	.4983	.0244	.6081	28
33	.21729	.97611	.22261	.4921	.0245	.6021	27
34	.21757	.97604	.22291	.4860	.0245	.5961	26
35	.21786	.97598	.22322	4.4799	1.0246	4.5901	25
36	.21814	.97592	.22353	.4737	.0247	.5841	24
37	.21843	.97585	.22383	.4676	.0247	.5782	23
38	.21871	.97579	.22414	.4615	.0248	.5722	22
39	.21899	.97573	.22444	.4555	.0249	.5663	21
40	.21928	.97566	.22475	4.4494	1.0249	4.5604	20
41	.21956	.97560	.22505	.4434	.0250	.5545	19
42	.21985	.97553	.22536	.4373	.0251	.5486	18
43	.22013	.97547	.22566	.4313	.0251	.5428	17
44	.22041	.97541	.22597	.4253	.0252	.5369	16
45	.22070	.97534	.22628	4.4194	1.0253	4.5311	15
46	.22098	.97528	.22658	.4134	.0253	.5253	14
47	.22126	.97521	.22689	.4074	.0254	.5195	13
48	.22155	.97515	.22719	.4015	.0255	.5137	12
49	.22183	.97508	.22750	.3956	.0255	.5079	11
50	.22211	.97502	.22781	4.3897	1.0256	4.5021	10
51	.22240	.97495	.22811	.3838	.0257	.4964	9
52	.22268	.97489	.22842	.3779	.0257	.4907	8
53	.22297	.97483	.22872	.3721	.0258	.4850	7
54	.22325	.97476	.22903	.3662	.0259	.4793	6
55	.22353	.97470	.22934	4.3604	1.0260	4.4736	5
56	.22382	.97463	.22964	.3546	.0260	.4679	4
57	.22410	.97457	.22995	.3488	.0261	.4623	3
58	.22438	.97450	.23025	.3430	.0262	.4566	2
59	.22467	.97443	.23056	.3372	.0262	.4510	1
60	.22495	.97437	.23087	4.3315	1.0263	4.4454	0
M	Cosine	Sine	Cotan.	Tan.	Cosec.	Secant	M

77°

111

<h2 style="text-align:center">13°</h2>

M	Sine	Cosine	Tan.	Cotan.	Secant	Cosec.	M
0	.22495	.97437	.23087	4.3315	1.0263	4.4454	60
1	.22523	.97430	.23117	.3257	.0264	.4398	59
2	.22552	.97424	.23148	.3200	.0264	.4342	58
3	.22580	.97417	.23179	.3143	.0265	.4287	57
4	.22608	.97411	.23209	.3086	.0266	.4231	56
5	.22637	.97404	.23240	4.3029	1.0266	4.4176	55
6	.22665	.97398	.23270	.2972	.0267	.4121	54
7	.22693	.97391	.23301	.2916	.0268	.4065	53
8	.22722	.97384	.23332	.2859	.0268	.4011	52
9	.22750	.97378	.23363	.2803	.0269	.3956	51
10	.22778	.97371	.23393	4.2747	1.0270	4.3901	50
11	.22807	.97364	.23424	.2691	.0271	.3847	49
12	.22835	.97358	.23455	.2635	.0271	3792	48
13	.22863	.97351	.23485	.2579	.0272	.3738	47
14	.22892	.97344	.23516	.2524	.0273	.3684	46
15	.22920	.97338	.23547	4.2468	1.0273	4.3630	45
16	.22948	.97331	.23577	.2413	.0274	.3576	44
17	.22977	.97324	.23608	.2358	.0275	.3522	43
18	.23005	.97318	.23639	.2303	.0276	.3469	42
19	.23033	.97311	.23670	.2248	.0276	.3415	41
20	.23061	.97304	.23700	4.2193	1.0277	4.3362	40
21	.23090	.97298	.23731	.2139	.0278	.3309	39
22	.23118	.97291	.23762	.2084	.0278	.3256	38
23	.23146	.97284	.23793	.2030	.0279	.3203	37
24	.23175	.97277	.23823	.1976	.0280	.3150	36
25	.23203	.97271	.23854	4.1921	1.0280	4.3098	35
26	.23231	.97264	.23885	.1867	.0281	.3045	34
27	.23260	.97257	.23916	.1814	.0282	.2993	33
28	.23288	.97250	.23946	.1760	.0283	.2941	32
29	.23316	.97244	.23977	.1706	.0283	.2888	31
30	.23344	.97237	.24008	4.1653	1.0284	4.2836	30
31	.23373	.97230	.24039	.1600	.0285	.2785	29
32	.23401	.97223	.24069	.1546	.0285	.2733	28
33	.23429	.97216	.24100	.1493	.0286	.2681	27
34	.23458	.97210	.24131	.1440	.0287	.2630	26
35	.23486	.97203	.24162	4.1388	1.0288	4.2579	25
36	.23514	.97196	.24192	.1335	.0288	.2527	24
37	.23542	.97189	.24223	.1282	.0289	.2476	23
38	.23571	.97182	.24254	.1230	.0290	.2425	22
39	.23599	.97175	.24285	.1178	.0291	.2375	21
40	.23627	.97169	.24316	4.1126	1.0291	4.2324	20
41	.23655	.97162	.24346	.1073	.0292	.2273	19
42	.23684	.97155	.24377	.1022	.0293	.2223	18
43	.23712	.97148	.24408	.0970	.0293	.2173	17
44	.23740	.97141	.24439	.0918	.0294	.2122	16
45	.23768	.97134	.24470	4.0867	1.0295	4.2072	15
46	.23797	.97127	.24501	.0815	.0296	.2022	14
47	.23825	.97120	.24531	.0764	.0296	.1972	13
48	.23853	.97113	.24562	.0713	.0297	.1923	12
49	.23881	.97106	.24593	.0662	.0298	.1873	11
50	.23910	.97099	.24624	4.0611	1.0299	4.1824	10
51	.23938	.97092	.24655	.0560	.0299	.1774	9
52	.23966	.97086	.24686	.0509	.0300	.1725	8
53	.23994	.97079	.24717	.0458	.0301	.1676	7
54	.24023	.97072	.24747	.0408	.0302	.1627	6
55	.24051	.97065	.24778	4.0358	1.0302	4.1578	5
56	.24079	.97058	.24809	.0307	.0303	.1529	4
57	.24107	.97051	.24840	.0257	.0304	.1481	3
58	.24136	.97044	.24871	.0207	.0305	.1432	2
59	.24164	.97037	.24902	.0157	.0305	.1384	1
60	.24192	.97029	.24933	4.0108	1.0306	4.1336	0
M	Cosine	Sine	Cotan.	Tan.	Cosec.	Secant	M

76°

<h2 style="text-align:center">14°</h2>

M	Sine	Cosine	Tan.	Cotan.	Secant	Cosec.	M
0	.24192	.97029	.24933	4.0108	1.0306	4.1336	60
1	.24220	.97022	.24964	.0058	.0307	.1287	59
2	.24249	.97015	.24995	.0009	.0308	.1239	58
3	.24277	.97008	.25025	3.9959	.0308	.1191	57
4	.24305	.97001	.25056	.9910	.0309	.1144	56
5	.24333	.96994	.25087	3.9861	1.0310	4.1096	55
6	.24361	.96987	.25118	.9812	.0311	.1048	54
7	.24390	.96980	.25149	.9763	.0311	.1001	53
8	.24418	.96973	.25180	.9714	.0312	.0953	52
9	.24446	.96966	.25211	.9665	.0313	.0906	51
10	.24474	.96959	.25242	3.9616	1.0314	4.0859	50
11	.24502	.96952	.25273	.9568	.0314	.0812	49
12	.24531	.96944	.25304	.9520	.0315	.0765	48
13	.24559	.96937	.25335	.9471	.0316	.0718	47
14	.24587	.96930	.25366	.9423	.0317	.0672	46
15	.24615	.96923	.25397	3.9375	1.0317	4.0625	45
16	.24643	.96916	.25428	.9327	.0318	.0579	44
17	.24672	.96909	.25459	.9279	.0319	.0532	43
18	.24700	.96901	.25490	.9231	.0320	.0486	42
19	.24728	.96894	.25521	.9184	.0320	.0440	41
20	.24756	.96887	.25552	3.9136	1.0321	4.0394	40
21	.24784	.96880	.25583	.9089	.0322	.0348	39
22	.24813	.96873	.25614	.9042	.0323	.0302	38
23	.24841	.96865	.25645	.8994	.0323	.0256	37
24	.24869	.96858	.25676	.8947	.0324	.0211	36
25	.24897	.96851	.25707	3.8900	1.0325	4.0165	35
26	.24925	.96844	.25738	.8853	.0326	.0120	34
27	.24953	.96836	.25769	.8807	.0327	.0074	33
28	.24982	.96829	.25800	.8760	.0327	.0029	32
29	.25010	.96822	.25831	.8713	.0328	3.9984	31
30	.25038	.96815	.25862	3.8667	1.0329	3.9939	30
31	.25066	.96807	.25893	.8621	.0330	.9894	29
32	.25094	.96800	.25924	.8574	.0330	.9850	28
33	.25122	.96793	.25955	.8528	.0331	.9805	27
34	.25151	.96785	.25986	.8482	.0332	.9760	26
35	.25179	.96778	.26017	3.8436	1.0333	3.9716	25
36	.25207	.96771	.26048	.8390	.0334	.9672	24
37	.25235	.96763	.26079	.8345	.0334	.9627	23
38	.25263	.96756	.26110	.8299	.0335	.9583	22
39	.25291	.96749	.26141	.8254	.0336	.9539	21
40	.25319	.96741	.26172	3.8208	1.0337	3.9495	20
41	.25348	.96734	.26203	.8163	.0338	.9451	19
42	.25376	.96727	.26234	.8118	.0338	.9408	18
43	.25404	.96719	.26266	.8073	.0339	.9364	17
44	.25432	.96712	.26297	.8027	.0340	.9320	16
45	.25460	.96704	.26328	3.7983	1.0341	3.9277	15
46	.25488	.96697	.26359	.7938	.0341	.9234	14
47	.25516	.96690	.26390	.7893	.0342	.9190	13
48	.25544	.96682	.26421	.7848	.0343	.9147	12
49	.25573	.96675	.26452	.7804	.0344	.9104	11
50	.25601	.96667	.26483	3.7759	1.0345	3.9061	10
51	.25629	.96660	.26514	.7715	.0345	.9018	9
52	.25657	.96652	.26546	.7671	.0346	.8976	8
53	.25685	.96645	.26577	.7627	.0347	.8933	7
54	.25713	.96638	.26608	.7583	.0348	.8890	6
55	.25741	.96630	.26639	3.7539	1.0349	3.8848	5
56	.25769	.96623	.26670	.7495	.0349	.8805	4
57	.25798	.96615	.26701	.7451	.0350	.8763	3
58	.25826	.96608	.26732	.7407	.0351	.8721	2
59	.25854	.96600	.26764	.7364	.0352	.8679	1
60	.25882	.96592	.26795	3.7320	1.0353	3.8637	0
M	Cosine	Sine	Cotan.	Tan.	Cosec.	Secant	M

75°

15°

M	Sine	Cosine	Tan.	Cotan.	Secant	Cosec.	M
0	.25882	.96592	.26795	3.7320	1.0353	3.8637	60
1	.25910	.96585	.26826	.7277	.0353	.8595	59
2	.25938	.96577	.26857	.7234	.0354	.8553	58
3	.25966	.96570	.26888	.7191	.0355	.8512	57
4	.25994	.96562	.26920	.7147	.0356	.8470	56
5	.26022	.96555	.26951	3.7104	1.0357	3.8428	55
6	.26050	.96547	.26982	.7062	.0358	.8387	54
7	.26078	.96540	.27013	.7019	.0358	.8346	53
8	.26107	.96532	.27044	.6976	.0359	.8304	52
9	.26135	.96524	.27076	.6933	.0360	.8263	51
10	.26163	.96517	.27107	3.6891	1.0361	3.8222	50
11	.26191	.96509	.27138	.6848	.0362	.8181	49
12	.26219	.96502	.27169	.6806	.0362	.8140	48
13	.26247	.96494	.27201	.6764	.0363	.8100	47
14	.26275	.96486	.27232	.6722	.0364	.8059	46
15	.26303	.96479	.27263	3.6679	1.0365	3.8018	45
16	.26331	.96471	.27294	.6637	.0366	.7978	44
17	.26359	.96463	.27326	.6596	.0367	.7937	43
18	.26387	.96456	.27357	.6554	.0367	.7897	42
19	.26415	.96448	.27388	.6512	.0368	.7857	41
20	.26443	.96440	.27419	3.6470	1.0369	3.7816	40
21	.26471	.96433	.27451	.6429	.0370	.7776	39
22	.26499	.96425	.27482	.6387	.0371	.7736	38
23	.26527	.96417	.27513	.6346	.0371	.7697	37
24	.26556	.96409	.27544	.6305	.0372	.7657	36
25	.26584	.96402	.27576	3.6263	1.0373	3.7617	35
26	.26612	.96394	.27607	.6222	.0374	.7577	34
27	.26640	.96386	.27638	.6181	.0375	.7538	33
28	.26668	.96378	.27670	.6140	.0376	.7498	32
29	.26696	.96371	.27701	.6100	.0376	.7459	31
30	.26724	.96363	.27732	3.6059	1.0377	3.7420	30
31	.26752	.96355	.27764	.6018	.0378	.7380	29
32	.26780	.96347	.27795	.5977	.0379	.7341	28
33	.26808	.96340	.27826	.5937	.0380	.7302	27
34	.26836	.96332	.27858	.5896	.0381	.7263	26
35	.26864	.96324	.27889	3.5856	1.0382	3.7224	25
36	.26892	.96316	.27920	.5816	.0382	.7186	24
37	.26920	.96308	.27952	.5776	.0383	.7147	23
38	.26948	.96301	.27983	.5736	.0384	.7108	22
39	.26976	.96293	.28014	.5696	.0385	.7070	21
40	.27004	.96285	.28046	3.5656	1.0386	3.7031	20
41	.27032	.96277	.28077	.5616	.0387	.6993	19
42	.27060	.96269	.28109	.5576	.0387	.6955	18
43	.27088	.96261	.28140	.5536	.0388	.6917	17
44	.27116	.96253	.28171	.5497	.0389	.6878	16
45	.27144	.96245	.28203	3.5457	1.0390	3.6840	15
46	.27172	.96238	.28234	.5418	.0391	.6802	14
47	.27200	.96230	.28266	.5378	.0392	.6765	13
48	.27228	.96222	.28297	.5339	.0393	.6727	12
49	.27256	.96214	.28328	.5300	.0393	.6689	11
50	.27284	.96206	.28360	3.5261	1.0394	3.6651	10
51	.27312	.96198	.28391	.5222	.0395	.6614	9
52	.27340	.96190	.28423	.5183	.0396	.6576	8
53	.27368	.96182	.28454	.5144	1.0397	.6539	7
54	.27396	.96174	.28486	.5105	.0398	.6502	6
55	.27424	.96166	.28517	3.5066	1.0399	3.6464	5
56	.27452	.96158	.28549	.5028	.0399	.6427	4
57	.27480	.96150	.28580	.4989	.0400	.6390	3
58	.27508	.96142	.28611	.4951	.0401	.6353	2
59	.27536	.96134	.28643	.4912	.0402	.6316	1
60	.27564	.96126	.28674	3.4874	1.0403	3.6279	0
M	Cosine	Sine	Cotan.	Tan.	Cosec.	Secant	M

74°

16°

M	Sine	Cosine	Tan.	Cotan.	Secant	Cosec.	M
0	.27564	.96126	.28674	3.4874	1.0403	3.6279	60
1	.27592	.96118	.28706	.4836	.0404	.6243	59
2	.27620	.96110	.28737	.4798	.0405	.6206	58
3	.27648	.96102	.28769	.4760	.0406	.6169	57
4	.27675	.96094	.28800	.4722	.0406	.6133	56
5	.27703	.96086	.28832	3.4684	1.0407	3.6096	55
6	.27731	.96078	.28863	.4646	.0408	.6060	54
7	.27759	.96070	.28895	.4608	.0409	.6024	53
8	.27787	.96062	.28926	.4570	.0410	.5987	52
9	.27815	.96054	.28958	.4533	.0411	.5951	51
10	.27843	.96045	.28990	3.4495	1.0412	3.5915	50
11	.27871	.96037	.29021	.4458	.0413	.5879	49
12	.27899	.96029	.29053	.4420	.0413	.5843	48
13	.27927	.96021	.29084	.4383	.0414	.5807	47
14	.27955	.96013	.29116	.4346	.0415	.5772	46
15	.27983	.96005	.29147	3.4308	1.0416	3.5736	45
16	.28011	.95997	.29179	.4271	.0417	.5700	44
17	.28039	.95989	.29210	.4234	.0418	.5665	43
18	.28067	.95980	.29242	.4197	.0419	.5629	42
19	.28094	.95972	.29274	.4160	.0420	.5594	41
20	.28122	.95964	.29305	3.4124	1.0420	3.5559	40
21	.28150	.95956	.29337	.4087	.0421	.5523	39
22	.28178	.95948	.29368	.4050	.0422	.5488	38
23	.28206	.95940	.29400	.4014	.0423	.5453	37
24	.28234	.95931	.29432	.3977	.0424	.5418	36
25	.28262	.95923	.29463	3.3941	1.0425	3.5383	35
26	.28290	.95915	.29495	.3904	.0426	.5348	34
27	.28318	.95907	.29526	.3868	.0427	.5313	33
28	.28346	.95898	.29558	.3832	.0428	.5279	32
29	.28374	.95890	.29590	.3795	.0428	.5244	31
30	.28401	.95882	.29621	3.3759	1.0429	3.5209	30
31	.28429	.95874	.29653	.3723	.0430	.5175	29
32	.28457	.95865	.29685	.3687	.0431	.5140	28
33	.28485	.95857	.29716	.3651	.0432	.5106	27
34	.28513	.95849	.29748	.3616	.0433	.5072	26
35	.28541	.95840	.29780	3.3580	1.0434	3.5037	25
36	.28569	.95832	.29811	.3544	.0435	.5003	24
37	.28597	.95824	.29843	.3509	.0436	.4969	23
38	.28624	.95816	.29875	.3473	.0437	3.4935	22
39	.28652	.95807	.29906	.3438	.0438	.4901	21
40	.28680	.95799	.29938	3.3402	1.0438	3.4867	20
41	.28708	.95791	.29970	.3367	.0439	.4833	19
42	.28736	.95782	.30001	.3332	.0440	.4799	18
43	.28764	.95774	.30033	.3296	.0441	.4766	17
44	.28792	.95765	.30065	.3261	.0442	.4732	16
45	.28820	.95757	.30096	3.3226	1.0443	3.4698	15
46	.28847	.95749	.30128	.3191	.0444	.4665	14
47	.28875	.95740	.30160	.3156	.0445	.4632	13
48	.28903	.95732	.30192	.3121	.0446	.4598	12
49	.28931	.95723	.30223	.3087	.0447	.4565	11
50	.28959	.95715	.30255	3.3052	1.0448	3.4532	10
51	.28987	.95707	.30287	.3017	.0448	.4498	9
52	.29014	.95698	.30319	3.2983	.0449	.4465	8
53	.29042	.95690	.30350	.2948	.0450	.4432	7
54	.29070	.95681	.30382	.2914	.0451	.4399	6
55	.29098	.95673	.30414	3.2879	1.0452	3.4366	5
56	.29126	.95664	.30446	.2845	.0453	.4334	4
57	.29154	.95656	.30478	.2811	.0454	.4301	3
58	.29181	.95647	.30509	.2777	.0455	.4268	2
59	.29209	.95639	.30541	.2742	.0456	.4236	1
60	.29237	.95630	.30573	3.2708	1.0457	3.4203	0
M	Cosine	Sine	Cotan.	Tan.	Cosec.	Secant	M

73°

M	Sine	Cosine	Tan.	Cotan.	Secant	Cosec.	M
0	.29237	.95630	.30573	3.2708	1.0457	3.4203	60
1	.29265	.95622	.30605	.2674	.0458	.4170	59
2	.29293	.95613	.30637	.2640	.0459	.4138	58
3	.29321	.95605	.30668	.2607	.0460	.4106	57
4	.29348	.95596	.30700	.2573	.0461	.4073	56
5	.29376	.95588	.30732	3.2539	1.0461	3.4041	55
6	.29404	.95579	.30764	.2505	.0462	.4009	54
7	.29432	.95571	.30796	.2472	.0463	.3977	53
8	.29460	.95562	.30828	.2438	.0464	.3945	52
9	.29487	.95554	.30859	.2405	.0465	.3913	51
10	.29515	.95545	.30891	3.2371	1.0466	3.3881	50
11	.29543	.95536	.30923	.2338	.0467	.3849	49
12	.29571	.95528	.30955	.2305	.0468	.3817	48
13	.29598	.95519	.30987	.2271	.0469	.3785	47
14	.29626	.95511	.31019	.2238	.0470	.3754	46
15	.29654	.95502	.31051	3.2205	1.0471	3.3722	45
16	.29682	.95493	.31083	.2172	.0472	.3690	44
17	.29710	.95485	.31115	.2139	.0473	.3659	43
18	.29737	.95476	.31146	.2106	.0474	.3627	42
19	.29765	.95467	.31178	.2073	.0475	.3596	41
20	.29793	.95459	.31210	3.2041	1.0476	3.3565	40
21	.29821	.95450	.31242	.2008	.0477	.3534	39
22	.29848	.95441	.31274	.1975	.0478	.3502	38
23	.29876	.95433	.31306	.1942	.0478	.3471	37
24	.29904	.95424	.31338	.1910	.0479	.3440	36
25	.29932	.95415	.31370	3.1877	1.0480	3.3409	35
26	.29959	.95407	.31402	.1845	.0481	.3378	34
27	.29987	.95398	.31434	.1813	.0482	.3347	33
28	.30015	.95389	.31466	.1780	.0483	.3316	32
29	.30043	.95380	.31498	.1748	.0484	.3286	31
30	.30070	.95372	.31530	3.1716	1.0485	3.3255	30
31	.30098	.95363	.31562	.1684	.0486	.3224	29
32	.30126	.95354	.31594	.1652	.0487	.3194	28
33	.30154	.95345	.31626	.1620	.0488	.3163	27
34	.30181	.95337	.31658	.1588	.0489	.3133	26
35	.30209	.95328	.31690	3.1556	1.0490	3.3102	25
36	.30237	.95319	.31722	.1524	.0491	.3072	24
37	.30265	.95310	.31754	.1492	.0492	.3042	23
38	.30292	.95301	.31786	.1460	.0493	.3011	22
39	.30320	.95293	.31818	.1429	.0494	.2981	21
40	.30348	.95284	.31850	3.1397	1.0495	3.2951	20
41	.30375	.95275	.31882	.1366	.0496	.2921	19
42	.30403	.95266	.31914	1334	.0497	.2891	18
43	.30431	.95257	.31946	.1303	.0498	.2861	17
44	.30459	.95248	.31978	1271	.0499	.2831	16
45	.30486	.95239	.32010	3.1240	1.0500	3.2801	15
46	.30514	.95231	.32042	1209	.0501	.2772	14
47	.30542	.95222	.32074	.1177	.0502	.2742	13
48	.30569	.95213	.32106	.1146	.0503	.2712	12
49	.30597	.95204	.32138	.1115	.0504	.2683	11
50	.30625	.95195	.32171	3.1084	1.0505	3.2653	10
51	.30653	.95186	.32203	.1053	.0506	.2624	9
52	.30680	.95177	.32235	.1022	.0507	.2594	8
53	.30708	.95168	.32267	.0991	.0508	.2565	7
54	.30736	.95159	.32299	.0960	.0509	.2535	6
55	.30763	.95150	.32331	3.0930	1.0510	3.2506	5
56	.30791	.95141	.32363	.0899	.0511	.2477	4
57	.30819	.95132	.32395	.0868	.0512	.2448	3
58	.30846	.95124	.32428	.0838	.0513	.2419	2
59	.30874	.95115	.32460	.0807	.0514	.2390	1
60	.30902	.95106	.32492	3.0777	1.0515	3.2361	0
M	Cosine	Sine	Cotan.	Tan.	Cosec.	Secant	M

M	Sine	Cosine	Tan.	Cotan.	Secant	Cosec.	M
0	.30902	.95106	.32492	3.0777	1.0515	3.2361	60
1	.30929	.95097	.32524	.0746	.0516	.2332	59
2	.30957	.95088	.32556	.0716	.0517	.2303	58
3	.30985	.95079	.32588	.0686	.0518	.2274	57
4	.31012	.95070	.32621	.0655	.0519	.2245	56
5	.31040	.95061	.32653	3.0625	1.0520	3.2216	55
6	.31068	.95051	.32685	.0595	.0521	.2188	54
7	.31095	.95042	.32717	.0565	.0522	.2159	53
8	.31123	.95033	.32749	.0535	.0523	.2131	52
9	.31150	.95024	.32782	.0505	.0524	.2102	51
10	.31178	.95015	.32814	3.0475	1.0525	3.2074	50
11	.31206	.95006	.32846	.0445	.0526	.2045	49
12	.31233	.94997	.32878	.0415	.0527	.2017	48
13	.31261	.94988	.32910	.0385	.0528	.1989	47
14	.31289	.94979	.32943	.0356	.0529	.1960	46
15	.31316	.94970	.32975	3.0326	1.0530	3.1932	45
16	.31344	.94961	.33007	.0296	.0531	.1904	44
17	.31372	.94952	.33039	.0267	.0532	.1876	43
18	.31399	.94942	.33072	.0237	.0533	.1848	42
19	.31427	.94933	.33104	.0208	.0534	.1820	41
20	.31454	.94924	.33136	3.0178	:.0535	3.1792	40
21	.31482	.94915	.33169	.0149	.0536	.1764	39
22	.31510	.94906	.33201	.0120	.0537	.1736	38
23	.31537	.94897	.33233	.0090	.0538	.1708	37
24	.31565	.94888	.33265	.0061	.0539	.1681	36
25	.31592	.94878	.33298	3.0032	1.0540	3.1653	35
26	.31620	.94869	.33330	.0003	.0541	.1625	34
27	.31648	.94860	.33362	2.9974	.0542	.1598	33
28	.31675	.94851	.33395	.9945	.0543	.1570	32
29	.31703	.94841	.33427	.9916	.0544	.1543	31
30	.31730	.94832	.33459	2.9887	1.0545	3.1515	30
31	.31758	.94823	.33492	.9858	.0546	.1488	29
32	.31786	.94814	.33524	.9829	.0547	.1461	28
33	.31813	.94805	.33557	.9800	.0548	.1433	27
34	.31841	.94795	.33589	.9772	.0549	.1406	26
35	.31868	.94786	.33621	2.9743	1.0550	3.1379	25
36	.31896	.94777	.33654	.9714	.0551	.1352	24
37	.31923	.94767	.33686	.9686	.0552	.1325	23
38	.31951	.94758	.33718	.9657	.0553	.1298	22
39	.31978	.94749	.33751	.9629	.0554	.1271	21
40	.32006	.94740	.33783	2.9600	1.0555	3.1244	20
41	.32034	.94730	.33816	.9572	.0556	.1217	19
42	.32061	.94721	.33848	.9544	.0557	.1190	18
43	.32089	.94712	.33880	.9515	.0558	.1163	17
44	.32116	.94702	.33913	.9487	.0559	.1137	16
45	.32144	.94693	.33945	2.9459	1.0560	3.1110	15
46	.32171	.94684	.33978	.9431	.0561	.1083	14
47	.32199	.94674	.34010	.9403	.0562	.1057	13
48	.32226	.94665	.34043	.9375	.0563	.1030	12
49	.32254	.94655	.34075	.9347	.0565	.1004	11
50	.32282	.94646	.34108	2.9319	1.0566	3.0977	10
51	.32309	.94637	.34140	.9291	.0567	.0951	9
52	.32337	.94627	.34173	.9263	.0568	.0925	8
53	.32364	.94618	.34205	.9235	.0569	.0898	7
54	.32392	.94608	.34238	.9208	.0570	.0872	6
55	.32419	.94599	.34270	2.9180	1.0571	3.0846	5
56	.32447	.94590	.34303	.9152	.0572	.0820	4
57	.32474	.94580	.34335	.9125	.0573	.0793	3
58	.32502	.94571	.34368	.9097	.0574	.0767	2
59	.32529	.94561	.34400	.9069	.0575	.0741	1
60	.32557	.94552	.34433	2.9042	1.0576	3.0715	0
M	Cosine	Sine	Cotan.	Tan.	Cosec.	Secant	M

M	Sine	Cosine	Tan.	Cotan.	Secant	Cosec.	M
0	.32557	.94552	.34433	2.9042	1.0576	3.0715	60
1	.32584	.94542	.34465	.9015	.0577	.0690	59
2	.32612	.94533	.34498	.8987	.0578	.0664	58
3	.32639	.94523	.34530	.8960	.0579	.0638	57
4	.32667	.94514	.34563	.8933	.0580	.0612	56
5	.32694	.94504	.34595	2.8905	1.0581	3.0586	55
6	.32722	.94495	.34628	.8878	.0582	.0561	54
7	.32749	.94485	.34661	.8851	.0584	.0535	53
8	.32777	.94476	.34693	.8824	.0585	.0509	52
9	.32804	.94466	.34726	.8797	.0586	.0484	51
10	.32832	.94457	.34758	2.8770	1.0587	3.0458	50
11	.32859	.94447	.34791	.8743	.0588	.0433	49
12	.32887	.94438	.34824	.8716	.0589	.0407	48
13	.32914	.94428	.34856	.8689	.0590	.0382	47
14	.32942	.94418	.34889	.8662	.0591	.0357	46
15	.32969	.94409	.34921	2.8636	1.0592	3.0331	45
16	.32996	.94399	.34954	.8609	.0593	.0306	44
17	.33024	.94390	.34987	.8582	.0594	.0281	43
18	.33051	.94380	.35019	.8555	.0595	.0256	42
19	.33079	.94370	.35052	.8529	.0596	.0231	41
20	.33106	.94361	.35085	2.8502	1.0598	3.0206	40
21	.33134	.94351	.35117	.8476	.0599	.0181	39
22	.33161	.94341	.35150	.8449	.0600	.0156	38
23	.33189	.94332	.35183	.8423	.0601	.0131	37
24	.33216	.94322	.35215	.8396	.0602	.0106	36
25	.33243	.94313	.35248	2.8370	1.0603	3.0081	35
26	.33271	.94303	.35281	.8344	.0604	.0056	34
27	.33298	.94293	.35314	.8318	.0605	.0031	33
28	.33326	.94283	.35346	.8291	.0606	.0007	32
29	.33353	.94274	.35379	.8265	.0607	2.9982	31
30	.33381	.94264	.35412	2.8239	1.0608	2.9957	30
31	.33408	.94254	.35445	.8213	.0609	.9933	29
32	.33435	.94245	.35477	.8187	.0611	.9908	28
33	.33463	.94235	.35510	.8161	.0612	.9884	27
34	.33490	.94225	.35543	.8185	.0613	.9859	26
35	.33518	.94215	.35576	2.8109	1.0614	2.9835	25
36	.33545	.94206	.35608	.8083	.0615	.9810	24
37	.33572	.94196	.35641	.8057	.0616	.9786	23
38	.33600	.94186	.35674	.8032	.0617	.9762	22
39	.33627	.94176	.35707	.8006	.0618	.9738	21
40	.33655	.94167	.35739	2.7980	1.0619	2.9713	20
41	.33682	.94157	.35772	.7954	.0620	.9689	19
42	.33709	.94147	.35805	.7929	.0622	.9665	18
43	.33737	.94137	.35838	.7903	.0623	.9641	17
44	.33764	.94127	.35871	.7878	.0624	.9617	16
45	.33792	.94118	.35904	2.7852	1.0625	2.9593	15
46	.33819	.94108	.35936	.7827	.0626	.9569	14
47	.33846	.94098	.35969	.7801	.0627	.9545	13
48	.33874	.94088	.36002	.7776	.0628	.9521	12
49	.33901	.94078	.36035	.7751	.0629	.9497	11
50	.33928	.94068	.36068	2.7725	1.0630	2.9474	10
51	.33956	.94058	.36101	.7700	.0632	.9450	9
52	.33983	.94049	.36134	.7675	.0633	.9426	8
53	.34011	.94039	.36167	.7650	.0634	.9402	7
54	.34038	.94029	.36199	.7625	.0635	.9379	6
55	.34065	.94019	.36232	2.7600	1.0636	2.9355	5
56	.34093	.94009	.36265	.7575	.0637	.9332	4
57	.34120	.93999	.36298	.7550	.0638	.9308	3
58	.34147	.93989	.36331	.7525	.0639	.9285	2
59	.34175	.93979	.36364	.7500	.0641	.9261	1
60	.34202	.93969	.36397	2.7475	1.0642	2.9238	0
M	Cosine	Sine	Cotan.	Tan.	Cosec.	Secant	M

M	Sine	Cosine	Tan.	Cotan.	Secant	Cosec.	M
0	.34202	.93969	.36397	2.7475	1.0642	2.9238	60
1	.34229	.93959	.36430	.7450	.0643	.9215	59
2	.34257	.93949	.36463	.7425	.0644	.9191	58
3	.34284	.93939	.36496	.7400	.0645	.9168	57
4	.34311	.93929	.36529	.7376	.0646	.9145	56
5	.34339	.93919	.36562	2.7351	1.0647	2.9122	55
6	.34366	.93909	.36595	.7326	.0648	.9098	54
7	.34393	.93899	.36628	.7302	.0650	.9075	53
8	.34421	.93889	.36661	.7277	.0651	.9052	52
9	.34448	.93879	.36694	.7252	.0652	.9029	51
10	.34475	.93869	.36727	2.7228	1.0653	2.9006	50
11	.34502	.93859	.36760	.7204	.0654	.8983	49
12	.34530	.93849	.36793	.7179	.0655	.8960	48
13	.34557	.93839	.36826	.7155	.0656	.8937	47
14	.34584	.93829	.36859	.7130	.0658	.8915	46
15	.34612	.93819	.36892	2.7106	1.0659	2.8892	45
16	.34639	.93809	.36925	.7082	.0660	.8869	44
17	.34666	.93799	.36958	.7058	.0661	.8846	43
18	.34693	.93789	.36991	.7033	.0662	.8824	42
19	.34721	.93779	.37024	.7009	.0663	.8801	41
20	.34748	.93769	.37057	2.6985	1.0664	2.8778	40
21	.34775	.93758	.37090	.6961	.0666	.8756	39
22	.34803	.93748	.37123	.6937	.0667	.8733	38
23	.34830	.93738	.37156	.6913	.0668	.8711	37
24	.34857	.93728	.37190	.6889	.0669	.8688	36
25	.34884	.93718	.37223	2.6865	1.0670	2.8666	35
26	.34912	.93708	.37256	.6841	.0671	.8644	34
27	.34939	.93698	.37289	.6817	.0673	.8621	33
28	.34966	.93687	.37322	.6794	.0674	.8599	32
29	.34993	.93677	.37355	.6770	.0675	.8577	31
30	.35021	.93667	.37388	2.6746	1.0676	2.8554	30
31	.35048	.93657	.37422	.6722	.0677	.8532	29
32	.35075	.93647	.37455	.6699	.0678	.8510	28
33	.35102	.93637	.37488	.6675	.0679	.8488	27
34	.35130	.93626	.37521	.6652	.0681	.8466	26
35	.35157	.93616	.37554	2.6628	1.0682	2.8444	25
36	.35184	.93606	.37587	.6604	.0683	.8422	24
37	.35211	.93596	.37621	.6581	.0684	.8400	23
38	.35239	.93585	.37654	.6558	.0685	.8378	22
39	.35266	.93575	.37687	.6534	.0686	.8356	21
40	.35293	.93565	.37720	2.6511	1.0688	2.8334	20
41	.35320	.93555	.37754	.6487	.0689	.8312	19
42	.35347	.93544	.37787	.6464	.0690	.8290	18
43	.35375	.93534	.37820	.6441	.0691	.8269	17
44	.35402	.93524	.37853	.6418	.0692	.8247	16
45	.35429	.93513	.37887	2.6394	1.0694	2.8225	15
46	.35456	.93503	.37920	.6371	.0695	.8204	14
47	.35483	.93493	.37953	.6348	.0696	.8182	13
48	.35511	.93482	.37986	.6325	.0697	.8160	12
49	.35538	.93472	.38020	.6302	.0698	.8139	11
50	.35565	.93462	.38053	2.6279	1.0699	2.8117	10
51	.35592	.93451	.38086	.6256	.0701	.8096	9
52	.35619	.93441	.38120	.6233	.0702	.8074	8
53	.35647	.93431	.38153	.6210	.0703	.8053	7
54	.35674	.93420	.38186	.6187	.0704	.8032	6
55	.35701	.93410	.38220	2.6164	1.0705	2.8010	5
56	.35728	.93400	.38253	.6142	.0707	.7989	4
57	.35755	.93389	.38286	.6119	.0708	.7968	3
58	.35782	.93379	.38320	.6096	.0709	.7947	2
59	.35810	.93368	.38353	.6073	.0710	.7925	1
60	.35837	.93358	.38386	2.6051	1.0711	2.7904	0
M	Cosine	Sine	Cotan.	Tan.	Cosec.	Secant	M

21°

M	Sine	Cosine	Tan.	Cotan.	Secant	Cosec.	M
0	.35837	.93358	.38386	2.6051	1.0711	2.7904	60
1	.35864	.93348	.38420	.6028	.0713	.7883	59
2	.35891	.93337	.38453	.6006	.0714	.7862	58
3	.35918	.93327	.38486	.5983	.0715	.7841	57
4	.35945	.93316	.38520	.5960	.0716	.7820	56
5	.35972	.93306	.38553	2.5938	1.0717	2.7799	55
6	.36000	.93295	.38587	.5916	.0719	.7778	54
7	.36027	.93285	.38620	.5893	.0720	.7757	53
8	.36054	.93274	.38654	.5871	.0721	.7736	52
9	.36081	.93264	.38687	.5848	.0722	.7715	51
10	.36108	.93253	.38720	2.5826	1.0723	2.7694	50
11	.36135	.93243	.38754	.5804	.0725	.7674	49
12	.36162	.93232	.38787	.5781	.0726	.7653	48
13	.36189	.93222	.38821	.5759	.0727	.7632	47
14	.36217	.93211	.38854	.5737	.0728	.7611	46
15	.36244	.93201	.38888	2.5715	1.0729	2.7591	45
16	.36271	.93190	.38921	.5693	.0731	.7570	44
17	.36298	.93180	.38955	.5671	.0732	.7550	43
18	.36325	.93169	.38988	.5649	.0733	.7529	42
19	.36352	.93158	.39022	.5627	.0734	.7509	41
20	.36379	.93148	.39055	2.5605	1.0736	2.7488	40
21	.36406	.93137	.39089	.5583	.0737	.7468	39
22	.36433	.93127	.39122	.5561	.0738	.7447	38
23	.36460	.93116	.39156	.5539	.0739	.7427	37
24	.36488	.93105	.39189	.5517	.0740	.7406	36
25	.36515	.93095	.39223	2.5495	1.0742	2.7386	35
26	.36542	.93084	.39257	.5473	.0743	.7366	34
27	.36569	.93074	.39290	.5451	.0744	.7346	33
28	.36596	.93063	.39324	.5430	.0745	.7325	32
29	.36623	.93052	.39357	.5408	.0747	.7305	31
30	.36650	.93042	.39391	2.5386	1.0748	2.7285	30
31	.36677	.93031	.39425	.5365	.0749	.7265	29
32	.36704	.93020	.39458	.5343	.0750	.7245	28
33	.36731	.93010	.39492	.5322	.0751	.7225	27
34	.36758	.92999	.39525	.5300	.0753	.7205	26
35	.36785	.92988	.39559	2.5278	1.0754	2.7185	25
36	.36812	.92978	.39593	.5257	.0755	.7165	24
37	.36839	.92967	.39626	.5236	.0756	.7145	23
38	.36866	.92956	.39660	.5214	.0758	.7125	22
39	.36893	.92945	.39694	.5193	.0759	.7105	21
40	.36921	.92935	.39727	2.5171	1.0760	2.7085	20
41	.36948	.92924	.39761	.5150	.0761	.7065	19
42	.36975	.92913	.39795	.5129	.0763	.7045	18
43	.37002	.92902	.39828	.5108	.0764	.7026	17
44	.37029	.92892	.39862	.5086	.0765	.7006	16
45	.37056	.92881	.39896	2.5065	1.0766	2.6986	15
46	.37083	.92870	.39930	.5044	.0768	.6967	14
47	.37110	.92859	.39963	.5023	.0769	.6947	13
48	.37137	.92848	.39997	.5002	.0770	.6927	12
49	.37164	.92838	.40031	.4981	.0771	.6908	11
50	.37191	.92827	.40065	2.4960	1.0773	2.6888	10
51	.37218	.92816	.40098	.4939	.0774	.6869	9
52	.37245	.92805	.40132	.4918	.0775	.6849	8
53	.37272	.92794	.40166	.4897	.0776	.6830	7
54	.37299	.92784	.40200	.4876	.0778	.6810	6
55	.37326	.92773	.40233	2.4855	1.0779	2.6791	5
56	.37353	.92762	.40267	.4834	.0780	.6772	4
57	.37380	.92751	.40301	.4813	.0781	.6752	3
58	.37407	.92740	.40335	.4792	.0783	.6733	2
59	.37434	.92729	.40369	.4772	.0784	.6714	1
60	.37461	.92718	.40403	2.4751	1.0785	2.6695	0
M	Cosine	Sine	Cotan.	Tan.	Cosec.	Secant	M

68°

22°

M	Sine	Cosine	Tan.	Cotan.	Secant	Cosec.	M
0	.37461	.92718	.40403	2.4751	1.0785	2.6695	60
1	.37488	.92707	.40436	.4730	.0787	.6675	59
2	.37514	.92696	.40470	.4709	.0788	.6656	58
3	.37541	.92686	.40504	.4689	.0789	.6637	57
4	.37568	.92675	.40538	.4668	.0790	.6618	56
5	.37595	.92664	.40572	2.4647	1.0792	2.6599	55
6	.37622	.92653	.40606	.4627	.0793	.6580	54
7	.37649	.92642	.40640	.4606	.0794	.6561	53
8	.37676	.92631	.40673	.4586	.0795	.6542	52
9	.37703	.92620	.40707	.4565	.0797	.6523	51
10	.37730	.92609	.40741	2.4545	1.0798	2.6504	50
11	.37757	.92598	.40775	.4525	.0799	.6485	49
12	.37784	.92587	.40809	.4504	.0801	.6466	48
13	.37811	.92576	.40843	.4484	.0802	.6447	47
14	.37838	.92565	.40877	.4463	.0803	.6428	46
15	.37865	.92554	.40911	2.4443	1.0804	2.6410	45
16	.37892	.92543	.40945	.4423	.0806	.6391	44
17	.37919	.92532	.40979	.4403	.0807	.6372	43
18	.37946	.92521	.41013	.4382	.0808	.6353	42
19	.37972	.92510	.41047	.4362	.0810	.6335	41
20	.37999	.92499	.41081	2.4342	1.0811	2.6316	40
21	.38026	.92488	.41115	.4322	.0812	.6297	39
22	.38053	.92477	.41149	.4302	.0813	.6279	38
23	.38080	.92466	.41183	.4282	.0815	.6260	37
24	.38107	.92455	.41217	.4262	.0816	.6242	36
25	.38134	.92443	.41251	2.4242	1.0817	2.6223	35
26	.38161	.92432	.41285	.4222	.0819	.6205	34
27	.38188	.92421	.41319	.4202	.0820	.6186	33
28	.38214	.92410	.41353	.4182	.0821	.6168	32
29	.38241	.92399	.41387	.4162	.0823	.6150	31
30	.38268	.92388	.41421	2.4142	1.0824	2.6131	30
31	.38295	.92377	.41455	.4122	.0825	.6113	29
32	.38322	.92366	.41489	.4102	.0826	.6095	28
33	.38349	.92354	.41524	.4083	.0828	.6076	27
34	.38376	.92343	.41558	.4063	.0829	.6058	26
35	.38403	.92332	.41592	2.4043	1.0830	2.6040	25
36	.38429	.92321	.41626	.4023	.0832	.6022	24
37	.38456	.92310	.41660	.4004	.0833	.6003	23
38	.38483	.92299	.41694	.3984	.0834	.5985	22
39	.38510	.92287	.41728	.3964	.0836	.5967	21
40	.38537	.92276	.41762	2.3945	1.0837	2.5949	20
41	.38564	.92265	.41797	.3925	.0838	.5931	19
42	.38591	.92254	.41831	.3906	.0840	.5913	18
43	.38617	.92242	.41865	.3886	.0841	.5895	17
44	.38644	.92231	.41899	.3867	.0842	.5877	16
45	.38671	.92220	.41933	2.3847	1.0844	2.5859	15
46	.38698	.92209	.41968	.3828	.0845	.5841	14
47	.38725	.92197	.42002	.3808	.0846	.5823	13
48	.38751	.92186	.42036	.3789	.0847	.5805	12
49	.38778	.92175	.42070	.3770	.0849	.5787	11
50	.38805	.92164	.42105	2.3750	1.0850	2.5770	10
51	.38832	.92152	.42139	.3731	.0851	.5752	9
52	.38859	.92141	.42173	.3712	.0853	.5734	8
53	.38886	.92130	.42207	.3692	.0854	.5716	7
54	.38912	.92118	.42242	.3673	.0855	.5699	6
55	.38939	.92107	.42276	2.3654	1.0857	2.5681	5
56	.38966	.92096	.42310	.3635	.0858	.5663	4
57	.38993	.92084	.42344	.3616	.0859	.5646	3
58	.39019	.92073	.42379	.3597	.0861	.5628	2
59	.39046	.92062	.42413	.3577	.0862	.5610	1
60	.39073	.92050	.42447	2.3558	1.0864	2.5593	0
M	Cosine	Sine	Cotan.	Tan.	Cosec.	Secant	M

67°

23°

M	Sine	Cosine	Tan.	Cotan.	Secant	Cosec.	M
0	.39073	.92050	.42447	2.3558	1.0864	2.5593	60
1	.39100	.92039	.42482	.3539	.0865	.5575	59
2	.39126	.92028	.42516	.3520	.0866	.5558	58
3	.39153	.92016	.42550	.3501	.0868	.5540	57
4	.39180	.92005	.42585	.3482	.0869	.5523	56
5	.39207	.91993	.42619	2.3463	1.0870	2.5506	55
6	.39234	.91982	.42654	.3445	.0872	.5488	54
7	.39260	.91971	.42688	.3426	.0873	.5471	53
8	.39287	.91959	.42722	.3407	.0874	.5453	52
9	.39314	.91948	.42757	.3388	.0876	.5436	51
10	.39341	.91936	.42791	2.3369	1.0877	2.5419	50
11	.39367	.91925	.42826	.3350	.0878	.5402	49
12	.39394	.91913	.42860	.3332	.0880	.5384	48
13	.39421	.91902	.42894	.3313	.0881	.5367	47
14	.39448	.91891	.42929	.3294	.0882	.5350	46
15	.39474	.91879	.42963	2.3276	1.0884	2.5333	45
16	.39501	.91868	.42998	.3257	.0885	.5316	44
17	.39528	.91856	.43032	.3238	.0886	.5299	43
18	.39554	.91845	.43067	.3220	.0888	.5281	42
19	.39581	.91833	.43101	.3201	.0889	.5264	41
20	.39608	.91822	.43136	2.3183	1.0891	2.5247	40
21	.39635	.91810	.43170	.3164	.0892	.5230	39
22	.39661	.91798	.43205	.3145	.0893	.5213	38
23	.39688	.91787	.43239	.3127	.0895	.5196	37
24	.39715	.91775	.43274	.3109	.0896	.5179	36
25	.39741	.91764	.43308	2.3090	1.0897	2.5163	35
26	.39768	.91752	.43343	.3072	.0899	.5146	34
27	.39795	.91741	.43377	.3053	.0900	.5129	33
28	.39821	.91729	.43412	.3035	.0902	.5112	32
29	.39848	.91718	.43447	.3017	.0903	.5095	31
30	.39875	.91706	.43481	2.2998	1.0904	2.5078	30
31	.39901	.91694	.43516	.2980	.0906	.5062	29
32	.39928	.91683	.43550	.2962	.0907	.5045	28
33	.39955	.91671	.43585	.2944	.0908	.5028	27
34	.39981	.91659	.43620	.2925	.0910	.5011	26
35	.40008	.91648	.43654	2.2907	1.0911	2.4995	25
36	.40035	.91636	.43689	.2889	.0913	.4978	24
37	.40061	.91625	.43723	.2871	.0914	.4961	23
38	.40088	.91613	.43758	.2853	.0915	.4945	22
39	.40115	.91601	.43793	.2835	.0917	.4928	21
40	.40141	.91590	.43827	2.2817	1.0918	2.4912	20
41	.40168	.91578	.43862	.2799	.0920	.4895	19
42	.40195	.91566	.43897	.2781	.0921	.4879	18
43	.40221	.91554	.43932	.2763	.0922	.4862	17
44	.40248	.91543	.43966	.2745	.0924	.4846	16
45	.40275	.91531	.44001	2.2727	1.0925	2.4829	15
46	.40301	.91519	.44036	.2709	.0927	.4813	14
47	.40328	.91508	.44070	.2691	.0928	.4797	13
48	.40354	.91496	.44105	.2673	.0929	.4780	12
49	.40381	.91484	.44140	.2655	.0931	.4764	11
50	.40408	.91472	.44175	2.2637	1.0932	2.4748	10
51	.40434	.91461	.44209	.2619	.0934	.4731	9
52	.40461	.91449	.44244	.2602	.0935	.4715	8
53	.40487	.91437	.44279	.2584	.0936	.4699	7
54	.40514	.91425	.44314	.2566	.0938	.4683	6
55	.40541	.91414	.44349	2.2548	1.0939	2.4666	5
56	.40567	.91402	.44383	.2531	.0941	.4650	4
57	.40594	.91390	.44418	.2513	.0942	.4634	3
58	.40620	.91378	.44453	.2495	.0943	.4618	2
59	.40647	.91366	.44488	.2478	.0945	.4602	1
60	.40674	.91354	.44523	2.2460	1.0946	2.4586	0
M	Cosine	Sine	Cotan.	Tan.	Cosec.	Secant	M

66°

24°

M	Sine	Cosine	Tan.	Cotan.	Secant	Cosec.	M
0	.40674	.91354	.44523	2.2460	1.0946	2.4586	60
1	.40700	.91343	.44558	.2443	.0948	.4570	59
2	.40727	.91331	.44593	.2425	.0949	.4554	58
3	.40753	.91319	.44627	.2408	.0951	.4538	57
4	.40780	.91307	.44662	.2390	.0952	.4522	56
5	.40806	.91295	.44697	2.2373	1.0953	2.4506	55
6	.40833	.91283	.44732	.2355	.0955	.4490	54
7	.40860	.91271	.44767	.2338	.0956	.4474	53
8	.40886	.91260	.44802	.2320	.0958	.4458	52
9	.40913	.91248	.44837	.2303	.0959	.4442	51
10	.40939	.91236	.44872	2.2286	1.0961	2.4426	50
11	.40966	.91224	.44907	.2268	.0962	.4411	49
12	.40992	.91212	.44942	.2251	.0963	.4395	48
13	.41019	.91200	.44977	.2234	.0965	.4379	47
14	.41045	.91188	.45012	.2216	.0966	.4363	46
15	.41072	.91176	.45047	2.2199	1.0968	2.4347	45
16	.41098	.91164	.45082	.2182	.0969	.4332	44
17	.41125	.91152	.45117	.2165	.0971	.4316	43
18	.41151	.91140	.45152	.2147	.0972	.4300	42
19	.41178	.91128	.45187	.2130	.0973	.4285	41
20	.41204	.91116	.45222	2.2113	1.0975	2.4269	40
21	.41231	.91104	.45257	.2096	.0976	.4254	39
22	.41257	.91092	.45292	.2079	.0978	.4238	38
23	.41284	.91080	.45327	.2062	.0979	.4222	37
24	.41310	.91068	.45362	.2045	.0981	.4207	36
25	.41337	.91056	.45397	2.2028	1.0982	2.4191	35
26	.41363	.91044	.45432	.2011	.0984	.4176	34
27	.41390	.91032	.45467	.1994	.0985	.4160	33
28	.41416	.91020	.45502	.1977	.0986	.4145	32
29	.41443	.91008	.45537	.1960	.0988	.4130	31
30	.41469	.90996	.45573	2.1943	1.0989	2.4114	30
31	.41496	.90984	.45608	.1926	.0991	.4099	29
32	.41522	.90972	.45643	.1909	.0992	.4083	28
33	.41549	.90960	.45678	.1892	.0994	.4068	27
34	.41575	.90948	.45713	.1875	.0995	.4053	26
35	.41602	.90936	.45748	2.1859	1.0997	2.4037	25
36	.41628	.90924	.45783	.1842	.0998	.4022	24
37	.41654	.90911	.45819	.1825	.1000	.4007	23
38	.41681	.90899	.45854	.1808	.1001	.3992	22
39	.41707	.90887	.45889	.1792	.1003	.3976	21
40	.41734	.90875	.45924	2.1775	1.1004	2.3961	20
41	.41760	.90863	.45960	.1758	.1005	.3946	19
42	.41787	.90851	.45995	.1741	.1007	.3931	18
43	.41813	.90839	.46030	.1725	.1008	.3916	17
44	.41839	.90826	.46065	.1708	.1010	.3901	16
45	.41866	.90814	.46101	2.1692	1.1011	2.3886	15
46	.41892	.90802	.46136	.1675	.1013	.3871	14
47	.41919	.90790	.46171	.1658	.1014	.3856	13
48	.41945	.90778	.46206	.1642	.1016	.3841	12
49	.41972	.90765	.46242	.1625	.1017	.3826	11
50	.41998	.90753	.46277	2.1609	1.1019	2.3811	10
51	.42024	.90741	.46312	.1592	.1020	.3796	9
52	.42051	.90729	.46348	.1576	.1022	.3781	8
53	.42077	.90717	.46383	.1559	.1023	.3766	7
54	.42103	.90704	.46418	.1543	.1025	.3751	6
55	.42130	.90692	.46454	2.1527	1.1026	2.3736	5
56	.42156	.90680	.46489	.1510	.1028	.3721	4
57	.42183	.90668	.46524	.1494	.1029	.3706	3
58	.42209	.90655	.46560	.1478	.1031	.3691	2
59	.42235	.90643	.46595	.1461	.1032	.3677	1
60	.42262	.90631	.46631	2.1445	1.1034	2.3662	0
M	Cosine	Sine	Cotan.	Tan.	Cosec.	Secant	M

65°

25°

M	Sine	Cosine	Tan.	Cotan.	Secant	Cosec.	M
0	.42262	.90631	.46631	2.1445	1.1034	2.3662	60
1	.42288	.90618	.46666	.1429	.1035	.3647	59
2	.42314	.90606	.46702	.1412	.1037	.3632	58
3	.42341	.90594	.46737	.1396	.1038	.3618	57
4	.42367	.90581	.46772	.1380	.1040	.3603	56
5	.42394	.90569	.46808	2.1364	1.1041	2.3588	55
6	.42420	.90557	.46843	.1348	.1043	.3574	54
7	.42446	.90544	.46879	.1331	.1044	.3559	53
8	.42473	.90532	.46914	.1315	.1046	.3544	52
9	.42499	.90520	.46950	.1299	.1047	.3530	51
10	.42525	.90507	.46985	2.1283	1.1049	2.3515	50
11	.42552	.90495	.47021	.1267	.1050	.3501	49
12	.42578	.90483	.47056	.1251	.1052	.3486	48
13	.42604	.90470	.47092	.1235	.1053	.3472	47
14	.42630	.90458	.47127	.1219	.1055	.3457	46
15	.42657	.90445	.47163	2.1203	1.1056	2.3443	45
16	.42683	.90433	.47199	.1187	.1058	.3428	44
17	.42709	.90421	.47234	.1171	.1059	.3414	43
18	.42736	.90408	.47270	.1155	.1061	.3399	42
19	.42762	.90396	.47305	.1139	.1062	.3385	41
20	.42788	.90383	.47341	2.1123	1.1064	2.3371	40
21	.42815	.90371	.47376	.1107	.1065	.3356	39
22	.42841	.90358	.47412	.1092	.1067	.3342	38
23	.42867	.90346	.47448	.1076	.1068	.3328	37
24	.42893	.90333	.47483	.1060	.1070	.3313	36
25	.42920	.90321	.47519	2.1044	1.1072	2.3299	35
26	.42946	.90308	.47555	.1028	.1073	.3285	34
27	.42972	.90296	.47590	.1013	.1075	.3271	33
28	.42998	.90283	.47626	.0997	.1076	.3256	32
29	.43025	.90271	.47662	.0981	.1078	.3242	31
30	.43051	.90258	.47697	2.0965	1.1079	2.3228	30
31	.43077	.90246	.47733	.0950	.1081	.3214	29
32	.43104	.90233	.47769	.0934	.1082	.3200	28
33	.43130	.90221	.47805	.0918	.1084	.3186	27
34	.43156	.90208	.47840	.0903	.1085	.3172	26
35	.43182	.90196	.47876	2.0887	1.1087	2.3158	25
36	.43208	.90183	.47912	.0872	.1088	.3143	24
37	.43235	.90171	.47948	.0856	.1090	.3129	23
38	.43261	.90158	.47983	.0840	.1092	.3115	22
39	.43287	.90145	.48019	.0825	.1093	.3101	21
40	.43313	.90133	.48055	2.0809	1.1095	2.3087	20
41	.43340	.90120	.48091	.0794	.1096	.3073	19
42	.43366	.90108	.48127	.0778	.1098	.3059	18
43	.43392	.90095	.48162	.0763	.1099	.3046	17
44	.43418	.90082	.48198	.0747	.1101	.3032	16
45	.43444	.90070	.48234	2.0732	1.1102	2.3018	15
46	.43471	.90057	.48270	.0717	.1104	.3004	14
47	.43497	.90044	.48306	.0701	.1106	.2990	13
48	.43523	.90032	.48342	.0686	.1107	.2976	12
49	.43549	.90019	.48378	.0671	.1109	.2962	11
50	.43575	.90006	.48414	2.0655	1.1110	2.2949	10
51	.43602	.89994	.48449	.0640	.1112	.2935	9
52	.43628	.89981	.48485	.0625	.1113	.2921	8
53	.43654	.89968	.48521	.0609	.1115	.2907	7
54	.43680	.89956	.48557	.0594	.1116	.2894	6
55	.43706	.89943	.48593	2.0579	1.1118	2.2880	5
56	.43732	.89930	.48629	.0564	.1120	.2866	4
57	.43759	.89918	.48665	.0548	.1121	.2853	3
58	.43785	.89905	.48701	.0533	.1123	.2839	2
59	.43811	.89892	.48737	.0518	.1124	.2825	1
60	.43837	.89879	.48773	2.0503	1.1126	2.2812	0
M	Cosine	Sine	Cotan.	Tan.	Cosec.	Secant	M

64°

26°

M	Sine	Cosine	Tan.	Cotan.	Secant	Cosec.	M
0	.43837	.89879	.48773	2.0503	1.1126	2.2812	60
1	.43863	.89867	.48809	.0488	.1127	.2798	59
2	.43889	.89854	.48845	.0473	.1129	.2784	58
3	.43915	.89841	.48881	.0458	.1131	.2771	57
4	.43942	.89828	.48917	.0443	.1132	.2757	56
5	.43968	.89815	.48953	2.0427	1.1134	2.2744	55
6	.43994	.89803	.48989	.0412	.1135	.2730	54
7	.44020	.89790	.49025	.0397	.1137	.2717	53
8	.44046	.89777	.49062	2.0382	.1139	.2703	52
9	.44072	.89764	.49098	.0367	.1140	.2690	51
10	.44098	.89751	.49134	2.0352	1.1142	2.2676	50
11	.44124	.89739	.49170	.0338	.1143	.2663	49
12	.44150	.89726	.49206	.0323	.1145	.2650	48
13	.44177	.89713	.49242	.0308	.1147	.2636	47
14	.44203	.89700	.49278	.0293	.1148	.2623	46
15	.44229	.89687	.49314	2.0278	1.1150	2.2610	45
16	.44255	.89674	.49351	.0263	.1151	.2596	44
17	.44281	.89661	.49387	.0248	.1153	.2583	43
18	.44307	.89649	.49423	.0233	.1155	.2570	42
19	.44333	.89636	.49459	.0219	.1156	.2556	41
20	.44359	.89623	.49495	2.0204	1.1158	2.2543	40
21	.44385	.89610	.49532	.0189	.1159	.2530	39
22	.44411	.89597	.49568	.0174	.1161	.2517	38
23	.44437	.89584	.49604	.0159	.1163	.2503	37
24	.44463	.89571	.49640	.0145	.1164	.2490	36
25	.44489	.89558	.49677	2.0130	1.1166	2.2477	35
26	.44516	.89545	.49713	.0115	.1167	.2464	34
27	.44542	.89532	.49749	.0101	.1169	.2451	33
28	.44568	.89519	.49785	.0086	.1171	.2438	32
29	.44594	.89506	.49822	.0071	.1172	.2425	31
30	.44620	.89493	.49858	2.0057	1.1174	2.2411	30
31	.44646	.89480	.49894	.0042	.1176	.2398	29
32	.44672	.89467	.49931	.0028	.1177	.2385	28
33	.44698	.89454	.49967	.0013	.1179	.2372	27
34	.44724	.89441	.50003	1.9998	.1180	.2359	26
35	.44750	.89428	.50040	1.9984	1.1182	2.2348	25
36	.44776	.89415	.50076	.9969	.1184	.2333	24
37	.44802	.89402	.50113	.9955	.1185	.2320	23
38	.44828	.89389	.50149	.9940	.1187	.2307	22
39	.44854	.89376	.50185	.9926	.1189	.2294	21
40	.44880	.89363	.50222	1.9912	1.1190	2.2282	20
41	.44906	.89350	.50258	.9897	.1192	.2269	19
42	.44932	.89337	.50295	.9883	.1193	.2256	18
43	.44958	.89324	.50331	.9868	.1195	.2243	17
44	.44984	.89311	.50368	.9854	.1197	.2230	16
45	.45010	.89298	.50404	1.9840	1.1198	2.2217	15
46	.45036	.89285	.50441	.9825	.1200	.2204	14
47	.45062	.89272	.50477	.9811	.1202	.2192	13
48	.45088	.89258	.50514	.9797	.1203	.2179	12
49	.45114	.89245	.50550	.9782	.1205	.2166	11
50	.45140	.89232	.50587	1.9768	1.1207	2.2153	10
51	.45166	.89219	.50623	.9754	.1208	.2141	9
52	.45191	.89206	.50660	.9739	.1210	.2128	8
53	.45217	.89193	.50696	.9725	.1212	.2115	7
54	.45243	.89180	.50733	.9711	.1213	.2103	6
55	.45269	.89166	.50769	1.9697	1.1215	2.2090	5
56	.45295	.89153	.50806	.9683	.1217	.2077	4
57	.45321	.89140	.50843	.9668	.1218	.2065	3
58	.45347	.89127	.50879	.9654	.1220	.2052	2
59	.45373	.89114	.50916	.9640	.1222	.2039	1
60	.45399	.89101	.50952	1.9626	1.1223	2.2027	0
M	Cosine	Sine	Cotan.	Tan.	Cosec.	Secaht	M

63°

27°

M	Sine	Cosine	Tan.	Cotan.	Secant	Cosec.	M
0	.45399	.89101	.50952	1.9626	1.1223	2.2027	60
1	.45425	.89087	.50989	.9612	.1225	.2014	59
2	.45451	.89074	.51026	.9598	.1226	.2002	58
3	.45477	.89061	.51062	.9584	.1228	.1989	57
4	.45503	.89048	.51099	.9570	.1230	.1977	56
5	.45528	.89034	.51136	1.9556	1.1231	2.1964	55
6	.45554	.89021	.51172	.9542	.1233	.1952	54
7	.45580	.89008	.51209	.9528	.1235	.1939	53
8	.45606	.88995	.51246	.9514	.1237	.1927	52
9	.45632	.88981	.51283	.9500	.1238	.1914	51
10	.45658	.88968	.51319	1.9486	1.1240	2.1902	50
11	.45684	.88955	.51356	.9472	.1242	.1889	49
12	.45710	.88942	.51393	.9458	.1243	.1877	48
13	.45736	.88928	.51430	.9444	.1245	.1865	47
14	.45761	.88915	.51466	.9430	.1247	.1852	46
15	.45787	.88902	.51503	1.9416	1.1248	2.1840	45
16	.45813	.88888	.51540	.9402	.1250	.1828	44
17	.45839	.88875	.51577	.9388	.1252	.1815	43
18	.45865	.88862	.51614	.9375	.1253	.1803	42
19	.45891	.88848	.51651	.9361	.1255	.1791	41
20	.45917	.88835	.51687	1.9347	1.1257	2.1778	40
21	.45942	.88822	.51724	.9333	.1258	.1766	39
22	.45968	.88808	.51761	.9319	.1260	.1754	38
23	.45994	.88795	.51798	.9306	.1262	.1742	37
24	.46020	.88781	.51835	.9292	.1264	.1730	36
25	.46046	.88768	.51872	1.9278	1.1265	2.1717	35
26	.46072	.88755	.51909	.9264	.1267	.1705	34
27	.46097	.88741	.51946	.9251	.1269	.1693	33
28	.46123	.88728	.51983	.9237	.1270	.1681	32
29	.46149	.88714	.52020	.9223	.1272	.1669	31
30	.46175	.88701	.52057	1.9210	1.1274	2.1657	30
31	.46201	.88688	.52094	.9196	.1275	.1645	29
32	.46226	.88674	.52131	.9182	.1277	.1633	28
33	.46252	.88661	.52168	.9169	.1279	.1620	27
34	.46278	.88647	.52205	.9155	.1281	.1608	26
35	.46304	.88634	.52242	1.9142	1.1282	2.1596	25
36	.46330	.88620	.52279	.9128	.1284	.1584	24
37	.46355	.88607	.52316	.9115	.1286	.1572	23
38	.46381	.88593	.52353	.9101	.1287	.1560	22
39	.46407	.88580	.52390	.9088	.1289	.1548	21
40	.46433	.88566	.52427	1.9074	1.1291	2.1536	20
41	.46458	.88553	.52464	.9061	.1293	.1525	19
42	.46484	.88539	.52501	.9047	.1294	.1513	18
43	.46510	.88526	.52538	.9034	.1296	.1501	17
44	.46536	.88512	.52575	.9020	.1298	.1489	16
45	.46561	.88499	.52612	1.9007	1.1299	2.1477	15
46	.46587	.88485	.52650	.8993	.1301	.1465	14
47	.46613	.88472	.52687	.8980	.1303	.1453	13
48	.46639	.88458	.52724	.8967	.1305	.1441	12
49	.46664	.88444	.52761	.8953	.1306	.1430	11
50	.46690	.88431	.52798	1.8940	1.1308	2.1418	10
51	.46716	.88417	.52836	.8927	.1310	.1406	9
52	.46741	.88404	.52873	.8913	.1312	.1394	8
53	.46767	.88390	.52910	.8900	.1313	.1382	7
54	.46793	.88376	.52947	.8887	.1315	.1371	6
55	.46819	.88363	.52984	1.8873	1.1317	2.1359	5
56	.46844	.88349	.53022	.8860	.1319	.1347	4
57	.46870	.88336	.53059	.8847	.1320	.1335	3
58	.46896	.88322	.53096	.8834	.1322	.1324	2
59	.46921	.88308	.53134	.8820	.1324	.1312	1
60	.46947	.88295	.53171	1.8807	1.1326	2.1300	0
M	Cosine	Sine	Cotan.	Tan.	Cosec.	Secant	M

62°

28°

M	Sine	Cosine	Tan.	Cotan.	Secant	Cosec.	M
0	.46947	.88295	.53171	1.8807	1.1326	2.1300	60
1	.46973	.88281	.53208	.8794	.1327	.1289	59
2	.46998	.88267	.53245	.8781	.1329	.1277	58
3	.47024	.88254	.53283	.8768	.1331	.1266	57
4	.47050	.88240	.53320	.8754	.1333	.1254	56
5	.47075	.88226	.53358	1.8741	1.1334	2.1242	55
6	.47101	.88213	.53395	.8728	.1336	.1231	54
7	.47127	.88199	.53432	.8715	.1338	.1219	53
8	.47152	.88185	.53470	.8702	.1340	.1208	52
9	.47178	.88171	.53507	.8689	.1341	.1196	51
10	.47204	.88158	.53545	1.8676	1.1343	2.1185	50
11	.47229	.88144	.53582	.8663	.1345	.1173	49
12	.47255	.88130	.53619	.8650	.1347	.1162	48
13	.47281	.88117	.53657	.8637	.1349	.1150	47
14	.47306	.88103	.53694	.8624	.1350	.1139	46
15	.47332	.88089	.53732	1.8611	1.1352	2.1127	45
16	.47357	.88075	.53769	.8598	.1354	.1116	44
17	.47383	.88061	.53807	.8585	.1356	.1104	43
18	.47409	.88048	.53844	.8572	.1357	.1093	42
19	.47434	.88034	.53882	.8559	.1359	.1082	41
20	.47460	.88020	.53919	1.8546	1.1361	2.1070	40
21	.47486	.88006	.53957	.8533	.1363	.1059	39
22	.47511	.87992	.53995	.8520	.1365	.1048	38
23	.47537	.87979	.54032	.8507	.1366	.1036	37
24	.47562	.87965	.54070	.8495	.1368	.1025	36
25	.47588	.87951	.54107	1.8482	1.1370	2.1014	35
26	.47613	.87937	.54145	.8469	.1372	.1002	34
27	.47639	.87923	.54183	.8456	.1373	.0991	33
28	.47665	.87909	.54220	.8443	.1375	.0980	32
29	.47690	.87895	.54258	.8430	.1377	.0969	31
30	.47716	.87882	.54295	1.8418	1.1379	2.0957	30
31	.47741	.87868	.54333	.8405	.1381	.0946	29
32	.47767	.87854	.54371	.8392	.1382	.0935	28
33	.47792	.87840	.54409	.8379	.1384	.0924	27
34	.47818	.87826	.54446	.8367	.1386	.0912	26
35	.47844	.87812	.54484	1.8354	1.1388	2.0901	25
36	.47869	.87798	.54522	.8341	.1390	.0890	24
37	.47895	.87784	.54559	.8329	.1391	.0879	23
38	.47920	.87770	.54597	.8316	.1393	.0868	22
39	.47946	.87756	.54635	.8303	.1395	.0857	21
40	.47971	.87742	.54673	1.8291	1.1397	2.0846	20
41	.47997	.87728	.54711	.8278	.1399	.0835	19
42	.48022	.87715	.54748	.8265	.1401	.0824	18
43	.48048	.87701	.54786	.8253	.1402	.0812	17
44	.48073	.87687	.54824	.8240	.1404	.0801	16
45	.48099	.87673	.54862	1.8227	1.1406	2.0790	15
46	.48124	.87659	.54900	.8215	.1408	.0779	14
47	.48150	.87645	.54937	.8202	.1410	.0768	13
48	.48175	.87631	.54975	.8190	.1411	.0757	12
49	.48201	.87617	.55013	.8177	.1413	.0746	11
50	.48226	.87603	.55051	1.8165	1.1415	2.0735	10
51	.48252	.87588	.55089	.8152	.1417	.0725	9
52	.48277	.87574	.55127	.8140	.1419	.0714	8
53	.48303	.87560	.55165	.8127	.1421	.0703	7
54	.48328	.87546	.55203	.8115	.1422	.0692	6
55	.48354	.87532	.55241	1.8102	1.1424	2.0681	5
56	.48379	.87518	.55279	.8090	.1426	.0670	4
57	.48405	.87504	.55317	.8078	.1428	.0659	3
58	.48430	.87490	.55355	.8065	.1430	.0648	2
59	.48455	.87476	.55393	.8053	.1432	.0637	1
60	.48481	.87462	.55431	1.8040	1.1433	2.0627	0
M	Cosine	Sine	Cotan.	Tan.	Cosec.	Secant	M

61°

29°

M	Sine	Cosine	Tan.	Cotan.	Secant	Cosec.	M
0	.48481	.87462	.55431	1.8040	1.1433	2.0627	60
1	.48506	.87448	.55469	.8028	.1435	.0616	59
2	.48532	.87434	.55507	.8016	.1437	.0605	58
3	.48557	.87420	.55545	.8003	.1439	.0594	57
4	.48583	.87405	.55583	.7991	.1441	.0583	56
5	.48608	.87391	.55621	1.7979	1.1443	2.0573	55
6	.48633	.87377	.55659	.7966	.1445	.0562	54
7	.48659	.87363	.55697	.7954	.1446	.0551	53
8	.48684	.87349	.55735	.7942	.1448	.0540	52
9	.48710	.87335	.55774	.7930	.1450	.0530	51
10	.48735	.87320	.55812	1.7917	1.1452	2.0519	50
11	.48760	.87306	.55850	.7905	.1454	.0508	49
12	.48786	.87292	.55888	.7893	.1456	.0498	48
13	.48811	.87278	.55926	.7881	.1458	.0487	47
14	.48837	.87264	.55964	.7868	.1459	.0476	46
15	.48862	.87250	.56003	1.7856	1.1461	2.0466	45
16	.48887	.87235	.56041	.7844	.1463	.0455	44
17	.48913	.87221	.56079	.7832	.1465	.0444	43
18	.48938	.87207	.56117	.7820	.1467	.0434	42
19	.48964	.87193	.56156	.7808	.1469	.0423	41
20	.48989	.87178	.56194	1.7795	1.1471	2.0413	40
21	.49014	.87164	.56232	.7783	.1473	.0402	39
22	.49040	.87150	.56270	.7771	.1474	.0392	38
23	.49065	.87136	.56309	.7759	.1476	.0381	37
24	.49090	.87121	.56347	.7747	.1478	.0370	36
25	.49116	.87107	.56385	1.7735	1.1480	2.0360	35
26	.49141	.87093	.56424	.7723	.1482	.0349	34
27	.49166	.87078	.56462	.7711	.1484	.0339	33
28	.49192	.87064	.56500	.7699	.1486	.0329	32
29	.49217	.87050	.56539	.7687	.1488	.0318	31
30	.49242	.87035	.56577	1.7675	1.1489	2.0308	30
31	.49268	.87021	.56616	.7663	.1491	.0297	29
32	.49293	.87007	.56654	.7651	.1493	.0287	28
33	.49318	.86992	.56692	.7639	.1495	.0276	27
34	.49343	.86978	.56731	.7627	.1497	.0266	26
35	.49369	.86964	.56769	1.7615	1.1499	2.0256	25
36	.49394	.86949	.56808	.7603	.1501	.0245	24
37	.49419	.86935	.56846	.7591	.1503	.0235	23
38	.49445	.86921	.56885	.7579	.1505	.0224	22
39	.49470	.86906	.56923	.7567	.1507	.0214	21
40	.49495	.86892	.56962	1.7555	1.1508	2.0204	20
41	.49521	.86877	.57000	.7544	.1510	.0194	19
42	.49546	.86863	.57039	.7532	.1512	.0183	18
43	.49571	.86849	.57077	.7520	.1514	.0173	17
44	.49596	.86834	.57116	.7508	.1516	.0163	16
45	.49622	.86820	.57155	1.7496	1.1518	2.0152	15
46	.49647	.86805	.57193	.7484	.1520	.0142	14
47	.49672	.86791	.57232	.7473	.1522	.0132	13
48	.49697	.86776	.57270	.7461	.1524	.0122	12
49	.49723	.86762	.57309	.7449	.1526	.0111	11
50	.49748	.86748	.57348	1.7437	1.1528	2.0101	10
51	.49773	.86733	.57386	.7426	.1530	.0091	9
52	.49798	.86719	.57425	.7414	.1531	.0081	8
53	.49823	.86704	.57464	.7402	.1533	.0071	7
54	.49849	.86690	.57502	.7390	.1535	.0061	6
55	.49874	.86675	.57541	1.7379	1.1537	2.0050	5
56	.49899	.86661	.57580	.7367	.1539	.0040	4
57	.49924	.86646	.57619	.7355	.1541	.0030	3
58	.49950	.86632	.57657	.7344	.1543	.0020	2
59	.49975	.86617	.57696	.7332	.1545	.0010	1
60	.50000	.86603	.57735	1.7320	1.1547	2.0000	0
M	Cosine	Sine	Cotan.	Tan.	Cosec.	Secant	M

60°

30°

M	Sine	Cosine	Tan.	Cotan.	Secant	Cosec.	M
0	.50000	.86603	.57735	1.7320	1.1547	2.0000	60
1	.50025	.86588	.57774	.7309	.1549	1.9990	59
2	.50050	.86573	.57813	.7297	.1551	.9980	58
3	.50075	.86559	.57851	.7286	.1553	.9970	57
4	.50101	.86544	.57890	.7274	.1555	.9960	56
5	.50126	.86530	.57929	1.7262	1.1557	1.9950	55
6	.50151	.86515	.57968	.7251	.1559	.9940	54
7	.50176	.86500	.58007	.7239	.1561	.9930	53
8	.50201	.86486	.58046	.7228	.1562	.9920	52
9	.50226	.86471	.58085	.7216	.1564	.9910	51
10	.50252	.86457	.58123	1.7205	1.1566	1.9900	50
11	.50277	.86442	.58162	.7193	.1568	.9890	49
12	.50302	.86427	.58201	.7182	.1570	.9880	48
13	.50327	.86413	.58240	.7170	.1572	.9870	47
14	.50352	.86398	.58279	.7159	.1574	.9860	46
15	.50377	.86383	.58318	1.7147	1.1576	1.9850	45
16	.50402	.86369	.58357	.7136	.1578	.9840	44
17	.50428	.86354	.58396	.7124	.1580	.9830	43
18	.50453	.86339	.58435	.7113	.1582	.9820	42
19	.50478	.86325	.58474	.7101	.1584	.9811	41
20	.50503	.86310	.58513	1.7090	1.1586	1.9801	40
21	.50528	.86295	.58552	.7079	.1588	.9791	39
22	.50553	.86281	.58591	.7067	.1590	.9781	38
23	.50578	.86266	.58630	.7056	.1592	.9771	37
24	.50603	.86251	.58670	.7044	.1594	.9761	36
25	.50628	.86237	.58709	1.7033	1.1596	1.9752	35
26	.50653	.86222	.58748	.7022	.1598	.9742	34
27	.50679	.86207	.58787	.7010	.1600	.9732	33
28	.50704	.86192	.58826	.6999	.1602	.9722	32
29	.50729	.86178	.58865	.6988	.1604	.9713	31
30	.50754	.86163	.58904	1.6977	1.1606	1.9703	30
31	.50779	.86148	.58944	.6965	.1608	.9693	29
32	.50804	.86133	.58983	.6954	.1610	.9683	28
33	.50829	.86118	.59022	.6943	.1612	.9674	27
34	.50854	.86104	.59061	.6931	.1614	.9664	26
35	.50879	.86089	.59100	1.6920	1.1616	1.9654	25
36	.50904	.86074	.59140	.6909	.1618	.9645	24
37	.50929	.86059	.59179	.6898	.1620	.9635	23
38	.50954	.86044	.59218	.6887	.1622	.9625	22
39	.50979	.86030	.59258	.6875	.1624	.9616	21
40	.51004	.86015	.59297	1.6864	1.1626	1.9606	20
41	.51029	.86000	.59336	.6853	.1628	.9596	19
42	.51054	.85985	.59376	.6842	.1630	9587	18
43	.51079	.85970	.59415	.6831	.1632	9577	17
44	.51104	.85955	.59454	.6820	.1634	.9568	16
45	.51129	.85941	.59494	1.6808	1.1636	1.9558	15
46	.51154	.85926	.59533	.6797	.1638	.9549	14
47	.51179	.85911	.59572	.6786	.1640	.9539	13
48	.51204	.85896	.59612	.6775	.1642	.9530	12
49	.51229	.85881	.59651	.6764	.1644	.9520	11
50	.51254	.85866	.59691	1.6753	1.1646	1.9510	10
51	.51279	.85851	.59730	.6742	.1648	.9501	9
52	.51304	.85836	.59770	.6731	.1650	.9491	8
53	.51329	.85821	.59809	.6720	.1652	.9482	7
54	.51354	.85806	.59849	.6709	.1654	.9473	6
55	.51379	.85791	.59888	1.6698	1.1656	1.9463	5
56	.51404	.85777	.59928	.6687	.1658	.9454	4
57	.51429	.85762	.59967	.6676	.1660	.9444	3
58	.51454	.85747	.60007	.6665	.1662	.9435	2
59	.51479	.85732	.60046	.6654	.1664	.9425	1
60	.51504	.85717	.60086	1.6643	1.1666	1.9416	0
M	Cosine	Sine	Cotan.	Tan.	Cosec.	Secant	M

59°

M	Sine	Cosine	Tan.	Cotan.	Secant	Cosec.	M
0	.51504	.85717	.60086	1.6643	1.1666	1.9416	60
1	.51529	.85702	.60126	.6632	.1668	.9407	59
2	.51554	.85687	.60165	.6621	.1670	.9397	58
3	.51578	.85672	.60205	.6610	.1672	.9388	57
4	.51603	.85657	.60244	.6599	.1674	.9378	56
5	.51628	.85642	.60284	1.6588	1.1676	1.9369	55
6	.51653	.85627	.60324	.6577	.1678	.9360	54
7	.51678	.85612	.60363	.6566	.1681	.9350	53
8	.51703	.85597	.60403	.6555	.1683	.9341	52
9	.51728	.85582	.60443	.6544	.1685	.9332	51
10	.51753	.85566	.60483	1.6534	1.1687	1.9322	50
11	.51778	.85551	.60522	.6523	.1689	.9313	49
12	.51803	.85536	.60562	.6512	.1691	.9304	48
13	.51827	.85521	.60602	.6501	.1693	.9295	47
14	.51852	.85506	.60642	.6490	.1695	.9285	46
15	.51877	.85491	.60681	1.6479	1.1697	1.9276	45
16	.51902	.85476	.60721	.6469	.1699	.9267	44
17	.51927	.85461	.60761	.6458	.1701	.9258	43
18	.51952	.85446	.60801	.6447	.1703	.9248	42
19	.51977	.85431	.60841	.6436	.1705	.9239	41
20	.52002	.85416	.60881	1.6425	1.1707	1.9230	40
21	.52026	.85400	.60920	.6415	.1709	.9221	39
22	.52051	.85385	.60960	.6404	.1712	.9212	38
23	.52076	.85370	.61000	.6393	.1714	.9203	37
24	.52101	.85355	.61040	.6383	.1716	.9193	36
25	.52126	.85340	.61080	1.6372	1.1718	1.9184	35
26	.52151	.85325	.61120	.6361	.1720	.9175	34
27	.52175	.85309	.61160	.6350	.1722	.9166	33
28	.52200	.85294	.61200	.6340	.1724	.9157	32
29	.52225	.85279	.61240	.6329	.1726	.9148	31
30	.52250	.85264	.61280	1.6318	1.1728	1.9139	30
31	.52275	.85249	.61320	.6308	.1730	.9130	29
32	.52299	.85234	.61360	.6297	.1732	.9121	28
33	.52324	.85218	.61400	.6286	.1734	.9112	27
34	.52349	.85203	.61440	.6276	.1737	.9102	26
35	.52374	.85188	.61480	1.6265	1.1739	1.9093	25
36	.52398	.85173	.61520	.6255	.1741	.9084	24
37	.52423	.85157	.61560	.6244	.1743	.9075	23
38	.52448	.85142	.61601	.6233	.1745	.9066	22
39	.52473	.85127	.61641	.6223	.1747	.9057	21
40	.52498	.85112	.61681	1.6212	1.1749	1.9048	20
41	.52522	.85096	.61721	.6202	.1751	.9039	19
42	.52547	.85081	.61761	.6191	.1753	.9030	18
43	.52572	.85066	.61801	.6181	.1756	.9021	17
44	.52597	.85050	.61842	.6170	.1758	.9013	16
45	.52621	.85035	.61882	1.6160	1.1760	1.9004	15
46	.52646	.85020	.61922	.6149	.1762	.8995	14
47	.52671	.85004	.61962	.6139	.1764	.8986	13
48	.52695	.84989	.62003	.6128	.1766	.8977	12
49	.52720	.84974	.62043	.6118	.1768	.8968	11
50	.52745	.84959	.62083	1.6107	1.1770	1.8959	10
51	.52770	.84943	.62123	.6097	.1772	.8950	9
52	.52794	.84928	.62164	.6086	.1775	.8941	8
53	.52819	.84912	.62204	.6076	.1777	.8932	7
54	.52844	.84897	.62244	.6066	.1779	.8924	6
55	.52868	.84882	.62285	1.6055	1.1781	1.8915	5
56	.52893	.84866	.62325	.6045	.1783	.8906	4
57	.52918	.84851	.62366	.6034	.1785	.8897	3
58	.52942	.84836	.62406	.6024	.1787	.8888	2
59	.52967	.84820	.62446	.6014	.1790	.8879	1
60	.52992	.84805	.62487	1.6003	1.1792	1.8871	0
M	Cosine	Sine	Cotan.	Tan.	Cosec.	Secant	M

M	Sine	Cosine	Tan.	Cotan.	Secant	Cosec.	M
0	.52992	.84805	.62487	1.6003	1.1792	1.8871	60
1	.53016	.84789	.62527	.5993	.1794	.8862	59
2	.53041	.84774	.62568	.5983	.1796	.8853	58
3	.53066	.84758	.62608	.5972	.1798	.8844	57
4	.53090	.84743	.62649	.5962	.1800	.8836	56
5	.53115	.84728	.62689	1.5952	1.1802	1.8827	55
6	.53140	.84712	.62730	.5941	.1805	.8818	54
7	.53164	.84697	.62770	.5931	.1807	.8809	53
8	.53189	.84681	.62811	.5921	.1809	.8801	52
9	.53214	.84666	.62851	.5910	.1811	.8792	51
10	.53238	.84650	.62892	1.5900	1.1813	1.8783	50
11	.53263	.84635	.62933	.5890	.1815	.8775	49
12	.53288	.84619	.62973	.5880	.1818	.8766	48
13	.53312	.84604	.63014	.5869	.1820	.8757	47
14	.53337	.84588	.63055	.5859	.1822	.8749	46
15	.53361	.84573	.63095	1.5849	1.1824	1.8740	45
16	.53386	.84557	.63136	.5839	.1826	.8731	44
17	.53411	.84542	.63177	.5829	.1828	.8723	43
18	.53435	.84526	.63217	.5818	.1831	.8714	42
19	.53460	.84511	.63258	.5808	.1833	.8706	41
20	.53484	.84495	.63299	1.5798	1.1835	1.8697	40
21	.53509	.84479	.63339	.5788	.1837	.8688	39
22	.53533	.84464	.63380	.5778	.1839	.8680	38
23	.53558	.84448	.63421	.5768	.1841	.8671	37
24	.53583	.84433	.63462	.5757	.1844	.8663	36
25	.53607	.84417	.63503	1.5747	1.1846	1.8654	35
26	.53632	.84402	.63543	.5737	.1848	.8646	34
27	.53656	.84386	.63584	.5727	.1850	.8637	33
28	.53681	.84370	.63625	.5717	.1852	.8629	32
29	.53705	.84355	.63666	.5707	.1855	.8620	31
30	.53730	.84339	.63707	1.5697	1.1857	1.8611	30
31	.53754	.84323	.63748	.5687	.1859	.8603	29
32	.53779	.84308	.63789	.5677	.1861	.8595	28
33	.53803	.84292	.63830	.5667	.1863	.8586	27
34	.53828	.84276	.63871	.5657	.1866	.8578	26
35	.53852	.84261	.63912	1.5646	1.1868	1.8569	25
36	.53877	.84245	.63953	.5636	.1870	.8561	24
37	.53901	.84229	.63994	.5626	.1872	.8552	23
38	.53926	.84214	.64035	.5616	.1874	.8544	22
39	.53950	.84198	.64076	.5606	.1877	.8535	21
40	.53975	.84182	.64117	1.5596	1.1879	1.8527	20
41	.53999	.84167	.64158	.5586	.1881	.8519	19
42	.54024	.84151	.64199	.5577	.1883	.8510	18
43	.54048	.84135	.64240	.5567	.1886	.8502	17
44	.54073	.84120	.64281	.5557	.1888	.8493	16
45	.54097	.84104	.64322	1.5547	1.1890	1.8485	15
46	.54122	.84088	.64363	.5537	.1892	.8477	14
47	.54146	.84072	.64404	.5527	.1894	.8468	13
48	.54171	.84057	.64446	.5517	.1897	.8460	12
49	.54195	.84041	.64487	.5507	.1899	.8452	11
50	.54220	.84025	.64528	1.5497	1.1901	1.8443	10
51	.54244	.84009	.64569	.5487	.1903	.8435	9
52	.54268	.83993	.64610	.5477	.1906	.8427	8
53	.54293	.83978	.64652	.5467	.1908	.8418	7
54	.54317	.83962	.64693	.5458	.1910	.8410	6
55	.54342	.83946	.64734	1.5448	1.1912	1.8402	5
56	.54366	.83930	.64775	.5438	.1915	.8394	4
57	.54391	.83914	.64817	.5428	.1917	.8385	3
58	.54415	.83899	.64858	.5418	.1919	.8377	2
59	.54439	.83883	.64899	.5408	.1921	.8369	1
60	.54464	.83867	.64941	1.5399	1.1924	1.8361	0
M	Cosine	Sine	Cotan.	Tan.	Cosec.	Secant	M

33°

M	Sine	Cosine	Tan.	Cotan.	Secant	Cosec.	M
0	.54464	.83867	.64941	1.5395	1.1924	1.8361	60
1	.54488	.83851	.64982	.5389	.1926	.8352	59
2	.54513	.83835	.65023	.5379	.1928	.8344	58
3	.54537	.83819	.65065	.5369	.1930	.8336	57
4	.54561	.83804	.65106	.5359	.1933	.8328	56
5	.54586	.83788	.65148	1.5350	1.1935	1.8320	55
6	.54610	.83772	.65189	.5340	.1937	.8311	54
7	.54634	.83756	.65231	.5330	.1939	.8303	53
8	.54659	.83740	.65272	.5320	.1942	.8295	52
9	.54683	.83724	.65314	.5311	.1944	.8287	51
10	.54708	.83708	.65355	1.5301	1.1946	1.8279	50
11	.54732	.83692	.65397	.5291	.1948	.8271	49
12	.54756	.83676	.65438	.5282	.1951	.8263	48
13	.54781	.83660	.65480	.5272	.1953	.8255	47
14	.54805	.83644	.65521	.5262	.1955	.8246	46
15	.54829	.83629	.65563	1.5252	1.1958	1.8238	45
16	.54854	.83613	.65604	.5243	.1960	.8230	44
17	.54878	.83597	.65646	.5233	.1962	.8222	43
18	.54902	.83581	.65688	.5223	.1964	.8214	42
19	.54926	.83565	.65729	.5214	.1967	.8206	41
20	.54951	.83549	.65771	1.5204	1.1969	1.8198	40
21	.54975	.83533	.65813	.5195	.1971	.8190	39
22	.54999	.83517	.65854	.5185	.1974	.8182	38
23	.55024	.83501	.65896	.5175	.1976	.8174	37
24	.55048	.83485	.65938	.5166	.1978	.8166	36
25	.55072	.83469	.65980	1.5156	1.1980	1.8158	35
26	.55097	.83453	.66021	.5147	.1983	.8150	34
27	.55121	.83437	.66063	.5137	.1985	.8142	33
28	.55145	.83421	.66105	.5127	.1987	.8134	32
29	.55169	.83405	.66147	.5118	.1990	.8126	31
30	.55194	.83388	.66188	1.5108	1.1992	1.8118	30
31	.55218	.83372	.66230	.5099	.1994	.8110	29
32	.55242	.83356	.66272	.5089	.1997	.8102	28
33	.55266	.83340	.66314	.5080	.1999	.8094	27
34	.55291	.83324	.66356	.5070	.2001	.8086	26
35	.55315	.83308	.66398	1.5061	1.2004	1.8078	25
36	.55339	.83292	.66440	.5051	.2006	.8070	24
37	.55363	.83276	.66482	.5042	.2008	.8062	23
38	.55388	.83260	.66524	.5032	.2010	.8054	22
39	.55412	.83244	.66566	.5023	.2013	.8047	21
40	.55436	.83228	.66608	1.5013	1.2015	1.8039	20
41	.55460	.83211	.66650	.5004	.2017	.8031	19
42	.55484	.83195	.66692	.4994	.2020	.8023	18
43	.55509	.83179	.66734	.4985	.2022	.8015	17
44	.55533	.83163	.66776	.4975	.2024	.8007	16
45	.55557	.83147	.66818	1.4966	1.2027	1.7999	15
46	.55581	.83131	.66860	.4957	.2029	.7992	14
47	.55605	.83115	.66902	.4947	.2031	.7984	13
48	.55629	.83098	.66944	.4938	.2034	.7976	12
49	.55654	.83082	.66986	.4928	.2036	.7968	11
50	.55678	.83066	.67028	1.4919	1.2039	1.7960	10
51	.55702	.83050	.67071	.4910	.2041	.7953	9
52	.55726	.83034	.67113	.4900	.2043	.7945	8
53	.55750	.83017	.67155	.4891	.2046	.7937	7
54	.55774	.83001	.67197	.4881	.2048	.7929	6
55	.55799	.82985	.67239	1.4872	1.2050	1.7921	5
56	.55823	.82969	.67282	.4863	.2053	.7914	4
57	.55847	.82952	.67324	.4853	.2055	.7906	3
58	.55871	.82936	.67366	.4844	.2057	.7898	2
59	.55895	.82920	.67408	.4835	.2060	.7891	1
60	.55919	.82904	.67451	1.4826	1.2062	1.7883	0
M	Cosine	Sine	Cotan.	Tan.	Cosec.	Secant	M

56°

34°

M	Sine	Cosine	Tan.	Cotan.	Secant	Cosec.	M
0	.55919	.82904	.67451	1.4826	1.2062	1.7883	60
1	.55943	.82887	.67493	.4816	.2064	.7875	59
2	.55967	.82871	.67535	.4807	.2067	.7867	58
3	.55992	.82855	.67578	.4798	.2069	.7860	57
4	.56016	.82839	.67620	.4788	.2072	.7852	56
5	.56040	.82822	.67663	1.4779	1.2074	1.7844	55
6	.56064	.82806	.67705	.4770	.2076	.7837	54
7	.56088	.82790	.67747	.4761	.2079	.7829	53
8	.56112	.82773	.67790	.4751	.2081	.7821	52
9	.56136	.82757	.67832	.4742	.2083	.7814	51
10	.56160	.82741	.67875	1.4733	1.2086	1.7806	50
11	.56184	.82724	.67917	.4724	.2088	.7798	49
12	.56208	.82708	.67960	.4714	.2091	.7791	48
13	.56232	.82692	.68002	.4705	.2093	.7783	47
14	.56256	.82675	.68045	.4696	.2095	.7776	46
15	.56280	.82659	.68087	1.4687	1.2098	1.7768	45
16	.56304	.82643	.68130	.4678	.2100	.7760	44
17	.56328	.82626	.68173	.4669	.2103	.7753	43
18	.56353	.82610	.68215	.4659	.2105	.7745	42
19	.56377	.82593	.68258	.4650	.2107	.7738	41
20	.56401	.82577	.68301	1.4641	1.2110	1.7730	40
21	.56425	.82561	.68343	.4632	.2112	.7723	39
22	.56449	.82544	.68386	.4623	.2115	.7715	38
23	.56473	.82528	.68429	.4614	.2117	.7708	37
24	.56497	.82511	.68471	.4605	.2119	.7700	36
25	.56521	.82495	.68514	1.4595	1.2122	1.7693	35
26	.56545	.82478	.68557	.4586	.2124	.7685	34
27	.56569	.82462	.68600	.4577	.2127	.7678	33
28	.56593	.82445	.68642	.4568	.2129	.7670	32
29	.56617	.82429	.68685	.4559	.2132	.7663	31
30	.56641	.82413	.68728	1.4550	1.2134	1.7655	30
31	.56664	.82396	.68771	.4541	.2136	.7648	29
32	.56688	.82380	.68814	.4532	.2139	.7640	28
33	.56712	.82363	.68857	.4523	.2141	.7633	27
34	.56736	.82347	.68899	.4514	.2144	.7625	26
35	.56760	.82330	.68942	1.4505	1.2146	1.7618	25
36	.56784	.82314	.68985	.4496	.2149	.7610	24
37	.56808	.82297	.69028	.4487	.2151	.7603	23
38	.56832	.82280	.69071	.4478	.2153	.7596	22
39	.56856	.82264	.69114	.4469	.2156	.7588	21
40	.56880	.82247	.69157	1.4460	1.2158	1.7581	20
41	.56904	.82231	.69200	.4451	.2161	.7573	19
42	.56928	.82214	.69243	.4442	.2163	.7566	18
43	.56952	.82198	.69286	.4433	.2166	.7559	17
44	.56976	.82181	.69329	.4424	.2168	.7551	16
45	.57000	.82165	.69372	1.4415	1.2171	1.7544	15
46	.57023	.82148	.69415	.4406	.2173	.7537	14
47	.57047	.82131	.69459	.4397	.2175	.7529	13
48	.57071	.82115	.69502	.4388	.2178	.7522	12
49	.57095	.82098	.69545	.4379	.2180	.7514	11
50	.57119	.82082	.69588	1.4370	1.2183	1.7507	10
51	.57143	.82065	.69631	.4361	.2185	.7500	9
52	.57167	.82048	.69674	.4352	.2188	.7493	8
53	.57191	.82032	.69718	.4343	.2190	.7485	7
54	.57214	.82015	.69761	.4335	.2193	.7478	6
55	.57238	.81998	.69804	1.4326	1.2195	1.7471	5
56	.57262	.81982	.69847	.4317	.2198	.7463	4
57	.57286	.81965	.69891	.4308	.2200	.7456	3
58	.57310	.81948	.69934	.4299	.2203	.7449	2
59	.57334	.81932	.69977	.4290	.2205	.7442	1
60	.57358	.81915	.70021	1.4281	1.2208	1.7434	0
M	Cosine	Sine	Cotan.	Tan.	Cosec.	Secant	M

55°

35°

M	Sine	Cosine	Tan.	Cotan.	Secant	Cosec.	M
0	.57358	.81915	.70021	1.4281	1.2208	1.7434	60
1	.57381	.81898	.70064	.4273	.2210	.7427	59
2	.57405	.81882	.70107	.4264	.2213	.7420	58
3	.57429	.81865	.70151	.4255	.2215	.7413	57
4	.57453	.81848	.70194	.4246	.2218	.7405	56
5	.57477	.81832	.70238	1.4237	1.2220	1.7398	55
6	.57500	.81815	.70281	.4228	.2223	.7391	54
7	.57524	.81798	.70325	.4220	.2225	.7384	53
8	.57548	.81781	.70368	.4211	.2228	.7377	52
9	.57572	.81765	.70412	.4202	.2230	.7369	51
10	.57596	.81748	.70455	1.4193	1.2233	1.7362	50
11	.57619	.81731	.70499	.4185	.2235	.7355	49
12	.57643	.81714	.70542	.4176	.2238	.7348	48
13	.57667	.81698	.70586	.4167	.2240	.7341	47
14	.57691	.81681	.70629	.4158	.2243	.7334	46
15	.57714	.81664	.70673	1.4150	1.2245	1.7327	45
16	.57738	.81647	.70717	.4141	.2248	.7319	44
17	.57762	.81630	.70760	.4132	.2250	.7312	43
18	.57786	.81614	.70804	.4123	.2253	.7305	42
19	.57809	.81597	.70848	.4115	.2255	.7298	41
20	.57833	.81580	.70891	1.4106	1.2258	1.7291	40
21	.57857	.81563	.70935	.4097	.2260	.7284	39
22	.57881	.81546	.70979	.4089	.2263	.7277	38
23	.57904	.81530	.71022	.4080	.2265	.7270	37
24	.57928	.81513	.71066	.4071	.2268	.7263	36
25	.57952	.81496	.71110	1.4063	1.2270	1.7256	35
26	.57975	.81479	.71154	.4054	.2273	.7249	34
27	.57999	.81462	.71198	.4045	.2276	.7242	33
28	.58023	.81445	.71241	1.4037	.2278	.7234	32
29	.58047	.81428	.71285	.4028	.2281	.7227	31
30	.58070	.81411	.71329	1.4019	1.2283	1.7220	30
31	.58094	.81395	.71373	.4011	.2286	.7213	29
32	.58118	.81378	.71417	.4002	.2288	.7206	28
33	.58141	.81361	.71461	.3994	.2291	.7199	27
34	.58165	.81344	.71505	.3985	.2293	.7192	26
35	.58189	.81327	.71549	1.3976	1.2296	1.7185	25
36	.58212	.81310	.71593	.3968	.2298	.7178	24
37	.58236	.81293	.71637	.3959	1.2301	.7171	23
38	.58259	.81276	.71681	.3951	.2304	.7164	22
39	.58283	.81259	.71725	.3942	.2306	.7157	21
40	.58307	.81242	.71769	1.3933	1.2309	1.7151	20
41	.58330	.81225	.71813	.3925	.2311	.7144	19
42	.58354	.81208	.71857	.3916	.2314	.7137	18
43	.58378	.81191	.71901	.3908	.2316	.7130	17
44	.58401	.81174	.71945	.3899	.2319	.7123	16
45	.58425	.81157	.71990	1.3891	1.2322	1.7116	15
46	.58448	.81140	.72034	.3882	.2324	.7109	14
47	.58472	.81123	.72078	.3874	.2327	.7102	13
48	.58496	.81106	.72122	.3865	.2329	.7095	12
49	.58519	.81089	.72166	.3857	.2332	.7088	11
50	.58543	.81072	.72211	1.3848	1.2335	1.7081	10
51	.58566	.81055	.72255	.3840	.2337	.7075	9
52	.58590	.81038	.72299	1.3831	.2340	.7068	8
53	.58614	.81021	.72344	.3823	.2342	.7061	7
54	.58637	.81004	.72388	.3814	.2345	.7054	6
55	.58661	.80987	.72432	1.3806	1.2348	1.7047	5
56	.58684	.80970	.72477	.3797	.2350	.7040	4
57	.58708	.80953	.72521	.3789	.2353	.7033	3
58	.58731	.80936	.72565	.3781	.2355	.7027	2
59	.58755	.80919	.72610	.3772	.2358	.7020	1
60	.58778	.80902	.72654	1.3764	1.2361	1.7013	0
M	Cosine	Sine	Cotan.	Tan.	Cosec.	Secant	M

54°

36°

M	Sine	Cosine	Tan.	Cotan.	Secant	Cosec.	M
0	.58778	.80902	.72654	1.3764	1.2361	1.7013	60
1	.58802	.80885	.72699	.3755	.2363	.7006	59
2	.58825	.80867	.72743	.3747	.2366	.6999	58
3	.58849	.80850	.72788	.3738	.2368	.6993	57
4	.58873	.80833	.72832	1.3730	.2371	.6986	56
5	.58896	.80816	.72877	1.3722	1.2374	1.6979	55
6	.58920	.80799	.72921	.3713	.2376	.6972	54
7	.58943	.80782	.72966	.3705	.2379	.6965	53
8	.58967	.80765	.73010	.3697	.2382	.6959	52
9	.58990	.80747	.73055	.3688	.2384	.6952	51
10	.59014	.80730	.73100	1.3680	1.2387	1.6945	50
11	.59037	.80713	.73144	.3672	.2389	.6938	49
12	.59060	.80696	.73189	.3663	.2392	.6932	48
13	.59084	.80679	.73234	.3655	.2395	.6925	47
14	.59107	.80662	.73278	.3647	.2397	.6918	46
15	.59131	.80644	.73323	1.3638	1.2400	1.6912	45
16	.59154	.80627	.73368	.3630	.2403	.6905	44
17	.59178	.80610	.73412	.3622	.2405	.6898	43
18	.59201	.80593	.73457	.3613	.2408	.6891	42
19	.59225	.80576	.73502	.3605	.2411	.6885	41
20	.59248	.80558	.73547	1.3597	1.2413	1.6878	40
21	.59272	.80541	.73592	.3588	.2416	.6871	39
22	.59295	.80524	.73637	.3580	.2419	.6865	38
23	.59318	.80507	.73681	.3572	.2421	.6858	37
24	.59342	.80489	.73726	.3564	.2424	.6851	36
25	.59365	.80472	.73771	1.3555	1.2427	1.6845	35
26	.59389	.80455	.73816	.3547	.2429	1.6838	34
27	.59412	.80437	.73861	.3539	.2432	.6831	33
28	.59435	.80420	.73906	.3531	.2435	.6825	32
29	.59459	.80403	.73951	.3522	.2437	.6818	31
30	.59482	.80386	.73996	1.3514	1.2440	1.6812	30
31	.59506	.80368	.74041	.3506	.2443	.6805	29
32	.59529	.80351	.74086	.3498	.2445	.6798	28
33	.59552	.80334	.74131	.3489	.2448	.6792	27
34	.59576	.80316	.74176	.3481	.2451	.6785	26
35	.59599	.80299	.74221	1.3473	1.2453	1.6779	25
36	.59622	.80282	.74266	.3465	.2456	.6772	24
37	.59646	.80264	.74312	.3457	.2459	.6766	23
38	.59669	.80247	.74357	.3449	.2461	.6759	22
39	.59692	.80230	.74402	.3440	.2464	.6752	21
40	.59716	.80212	.74447	1.3432	1.2467	1.6746	20
41	.59739	.80195	.74492	.3424	.2470	.6739	19
42	.59762	.80177	.74538	.3416	.2472	.6733	18
43	.59786	.80160	.74583	.3408	.2475	.6726	17
44	.59809	.80143	.74628	.3400	.2478	.6720	16
45	.59832	.80125	.74673	1.3392	1.2480	1.6713	15
46	.59856	.80108	.74719	1.3383	.2483	.6707	14
47	.59879	.80090	.74764	.3375	.2486	.6700	13
48	.59902	.80073	.74809	.3367	.2488	.6694	12
49	.59926	.80056	.74855	.3359	.2491	.6687	11
50	.59949	.80038	.74900	1.3351	1.2494	1.6681	10
51	.59972	.80021	.74946	.3343	.2497	.6674	9
52	.59995	.80003	.74991	.3335	.2499	.6668	8
53	.60019	.79986	.75037	.3327	.2502	.6661	7
54	.60042	.79968	.75082	.3319	.2505	.6655	6
55	.60065	.79951	.75128	1.3311	1.2508	1.6648	5
56	.60088	.79933	.75173	.3303	.2510	.6642	4
57	.60112	.79916	.75219	.3294	.2513	.6636	3
58	.60135	.79898	.75264	.3286	.2516	.6629	2
59	.60158	.79881	.75310	.3278	.2519	.6623	1
60	.60181	.79863	.75355	1.3270	1.2521	1.6616	0
M	Cosine	Sine	Cotan.	Tan.	Cosec.	Secant	M

53°

37°

M	Sine	Cosine	Tan.	Cotan.	Secant	Cosec.	M
0	.60181	.79863	.75355	1.3270	1.2521	1.6616	60
1	.60205	.79346	.75401	.3262	.2524	.6610	59
2	.60228	.79828	.75447	.3254	.2527	.6603	58
3	.60251	.79811	.75492	.3246	.2530	.6597	57
4	.60274	.79793	.75538	.3238	.2532	.6591	56
5	.60298	.79776	.75584	1.3230	1.2535	1.6584	55
6	.60320	.79758	.75629	.3222	.2538	.6578	54
7	.60344	.79741	.75675	.3214	.2541	.6572	53
8	.60367	.79723	.75721	.3206	.2543	.6565	52
9	.60390	.79706	.75767	.3198	.2546	.6559	51
10	.60413	.79688	.75812	1.3190	1.2549	1.6552	50
11	.60437	.79670	.75858	.3182	.2552	.6546	49
12	.60460	.79653	.75904	.3174	.2554	.6540	48
13	.60483	.79635	.75950	.3166	.2557	.6533	47
14	.60506	.79618	.75996	.3159	.2560	.6527	46
15	.60529	.79600	.76042	1.3151	1.2563	1.6521	45
16	.60552	.79582	76088	.3143	.2565	.6514	44
17	.60576	.79565	.76134	.3135	.2568	.6508	43
18	.60599	.79547	.76179	.3127	.2571	.6502	42
19	.60622	.79530	.76225	.3119	.2574	.6496	41
20	.60645	.79512	.76271	1.3111	1.2577	1.6489	40
21	.60668	.79494	.76317	.3103	.2579	.6483	39
22	.60691	.79477	.76364	.3095	.2582	.6477	38
23	.60714	.79459	.76410	.3087	.2585	.6470	37
24	.60737	.79441	.76456	.3079	.2588	.6464	36
25	.60761	.79424	.76502	1.3071	1.2591	1.6458	35
26	.60784	.79406	.76548	.3064	.2593	.6452	34
27	.60807	.79388	.76594	.3056	.2596	.6445	33
28	.60830	.79371	.76640	.3048	.2599	.6439	32
29	.60853	.79353	.76686	.3040	.2602	.6433	31
30	.60876	.79335	.76733	1.3032	1.2605	1.6427	30
31	.60899	.79318	.76779	.3024	.2607	.6420	29
32	.60922	.79300	.76825	.3016	.2610	.6414	28
33	.60945	.79282	.76871	.3009	.2613	.6408	27
34	.60968	.79264	.76918	.3001	.2616	.6402	26
35	.60991	.79247	.76964	1.2993	1.2619	1.6396	25
36	.61014	.79229	.77010	.2985	.2622	1.6389	24
37	.61037	.79211	.77057	.2977	.2624	1.6383	23
38	.61061	.79193	77103	.2970	.2627	.6377	22
39	.61084	.79176	.77149	.2962	.2630	.6371	21
40	.61107	.79158	.77196	1.2954	1.2633	1.6365	20
41	.61130	.79140	.77242	.2946	.2636	.6359	19
42	.61153	.79122	.77289	.2938	.2639	.6352	18
43	.61176	.79104	.77335	.2931	.2641	.6346	17
44	.61199	.79087	.77382	.2923	.2644	.6340	16
45	.61222	.79069	.77428	1.2915	1.2647	1.6334	15
46	.61245	.79051	77475	.2907	.2650	.6328	14
47	.61268	.79033	.77521	.2900	.2653	.6322	13
48	.61290	.79015	.77568	.2892	.2656	.6316	12
49	.61314	.78998	.77614	.2884	.2659	.6309	11
50	.61337	.78980	.77661	1.2876	1.2661	1.6303	10
51	.61360	.78962	.77708	.2869	.2664	.6297	9
52	.61383	.78944	.77754	.2861	.2667	.6291	8
53	.61405	.78926	.77801	.2853	.2670	.6285	7
54	.61428	.78908	.77848	.2845	.2673	.6279	6
55	.61451	.78890	.77895	1.2838	1.2676	1.6273	5
56	.61474	.78873	.77941	.2830	.2679	.6267	4
57	.61497	.78855	.77988	.2822	.2681	.6261	3
58	.61520	.78837	.78035	.2815	.2684	.6255	2
59	.61543	.78819	.78082	.2807	.2687	.6249	1
60	.61566	.78801	.78128	1.2799	1.2690	1.6243	0
M	Cosine	Sine	Cotan.	Tan	Cosec.	Secant	M

52°

38°

M	Sine	Cosine	Tan.	Cotan.	Secant	Cosec.	M
0	.61566	.78801	.78128	1.2799	1.2690	1.6243	60
1	.61589	.78783	.78175	.2792	.2693	.6237	59
2	.61612	.78765	.78222	.2784	.2696	.6231	58
3	.61635	.78747	.78269	.2776	.2699	.6224	57
4	.61658	.78729	.78316	.2769	.2702	.6218	56
5	.61681	.78711	.78363	1.2761	1.2705	1.6212	55
6	.61703	.78693	.78410	.2753	.2707	.6206	54
7	.61726	.78675	.78457	.2746	.2710	.6200	53
8	.61749	.78657	.78504	.2738	.2713	.6194	52
9	.61772	.78640	.78551	.2730	.2716	.6188	51
10	.61795	.78622	.78598	1.2723	1.2719	1.6182	50
11	.61818	.78604	.78645	.2715	.2722	.6176	49
12	.61841	.78586	.78692	.2708	.2725	.6170	48
13	.61864	.78568	.78739	.2700	.2728	.6164	47
14	.61886	.78550	.78786	.2692	.2731	.6159	46
15	.61909	.78532	.78834	1.2685	1.2734	1.6153	45
16	.61932	.78514	.78881	.2677	.2737	.6147	44
17	.61955	.78496	.78928	.2670	.2739	.6141	43
18	.61978	78478	.78975	.2662	.2742	.6135	42
19	.62001	.78460	.79022	.2655	.2745	.6129	41
20	.62023	.78441	.79070	1.2647	1.2748	1.6123	40
21	.62046	.78423	.79117	.2639	.2751	.6117	39
22	.62069	.78405	.79164	.2632	.2754	.6111	38
23	.62092	.78387	.79212	.2624	.2757	.6105	37
24	.62115	.78369	.79259	.2617	.2760	.6099	36
25	.62137	.78351	.79306	1.2609	1.2763	1.6093	35
26	.62160	.78333	.79354	.2602	.2766	.6087	34
27	.62183	.78315	79401	.2594	.2769	.6081	33
28	.62206	.78297	.79449	.2587	.2772	.6077	32
29	.62229	.78279	.79496	.2579	.2775	.6070	31
30	.62251	.78261	.79543	1.2572	1.2778	1.6064	30
31	.62274	.78243	.79591	.2564	.2781	.6058	29
32	.62297	78224	.79639	.2557	.2784	.6052	28
33	.62320	.78206	.79686	.2549	.2787	.6046	27
34	.62342	.78188	.79734	.2542	.2790	.6040	26
35	.62365	.78170	.79781	1.2534	1.2793	1.6034	25
36	.62388	.78152	.79829	.2527	.2795	.6029	24
37	.62411	78134	.79876	.2519	.2798	.6023	23
38	.62433	.78116	.79924	.2512	.2801	.6017	22
39	.62456	.78097	.79972	.2504	.2804	.6011	21
40	.62479	.78079	.80020	1.2497	1.2807	1.6005	20
41	.62501	.78061	.80067	.2489	.2810	.6000	19
42	.62524	.78043	.80115	.2482	.2813	.5994	18
43	.62547	.78025	.80163	.2475	.2816	·.5988	17
44	.62570	78007	.80211	.2467	.2819	.5982	16
45	.62592	.77988	.80258	1.2460	1.2822	1.5976	15
46	.62615	.77970	.80306	.2452	.2825	.5971	14
47	.62638	.77952	.80354	.2445	.2828	.5965	13
48	.62660	.77934	.80402	.2437	.2831	.5959	12
49	.62683	.77915	.80450	.2430	.2834	.5953	11
50	.62706	.77897	.80498	1.2423	1.2837	1.5947	10
51	.62728	.77879	.80546	.2415	.2840	.5942	9
52	.62751	.77861	.80594	.2408	.2843	.5936	8
53	.62774	.77842	.80642	.2400	.2846	.5930	7
54	.62796	.77824	.80690	.2393	.2849	.5924	6
55	.62819	.77806	.80738	1.2386	1.2852	1.5919	5
56	.62841	.77788	.80786	.2378	.2855	.5913	4
57	.62854	.77769	.80834	.2371	.2858	.5907	3
58	.62887	.77751	.80882	.2364	.2861	.5901	2
59	.62909	.77733	.80930	.2356	.2864	.5896	1
60	.62932	.77715	.80978	1.2349	1.2867	1.5890	0
M	Cosine	Sine	Cotan.	Tan.	Cosec.	Secant	M

51°

M	Sine	Cosine	Tan.	Cotan.	Secant	Cosec.	M
0	.62932	.77715	.80978	1.2349	1.2867	1.5890	60
1	.62955	.77696	.81026	.2342	.2871	.5884	59
2	.62977	.77678	.81075	.2334	.2874	.5879	58
3	.63000	.77660	.81123	.2327	.2877	.5873	57
4	.63022	.77641	.81171	.2320	.2880	.5867	56
5	.63045	.77623	.81219	1.2312	1.2883	1.5862	55
6	.63067	.77605	.81268	.2305	.2886	.5856	54
7	.63090	.77586	.81316	.2297	.2889	.5850	53
8	.63113	.77568	.81364	.2290	.2892	.5845	52
9	.63135	.77549	.81413	.2283	.2895	.5839	51
10	.63158	.77531	.81461	1.2276	1.2898	1.5833	50
11	.63180	.77513	.81509	.2268	.2901	.5828	49
12	.63203	.77494	.81558	.2261	.2904	.5822	48
13	.63225	.77476	.81606	.2254	.2907	.5816	47
14	.63248	.77458	.81655	.2247	.2910	.5811	46
15	.63270	.77439	.81703	1.2239	1.2913	1.5805	45
16	.63293	.77421	.81752	.2232	.2916	.5799	44
17	.63315	.77402	.81800	.2225	.2919	.5794	43
18	.63338	.77384	.81849	.2218	.2922	.5788	42
19	.63360	.77365	.81898	.2210	.2926	.5783	41
20	.63383	.77347	.81946	1.2203	1.2929	1.5777	40
21	.63405	.77329	.81995	.2196	.2932	.5771	39
22	.63428	.77310	.82043	.2189	.2935	.5766	38
23	.63450	.77292	.82092	.2181	.2938	.5760	37
24	.63473	.77273	.82141	.2174	.2941	.5755	36
25	.63495	.77255	.82190	1.2167	1.2944	1.5749	35
26	.63518	.77236	.82238	.2160	.2947	.5743	34
27	.63540	.77218	.82287	.2152	.2950	.5738	33
28	.63563	.77199	.82336	.2145	.2953	.5732	32
29	.63585	.77181	.82385	.2138	.2956	.5727	31
30	.63608	.77162	.82434	1.2131	1.2960	1.5721	30
31	.63630	.77144	.82482	.2124	.2963	.5716	29
32	.63653	.77125	.82531	.2117	.2966	.5710	28
33	.63675	.77107	.82580	.2109	.2969	.5705	27
34	.63697	.77088	.82629	.2102	.2972	.5699	26
35	.63720	.77070	.82678	1.2095	1.2975	1.5694	25
36	.63742	.77051	.82727	.2088	.2978	.5688	24
37	.63765	.77033	.82776	.2081	.2981	.5683	23
38	.63787	.77014	.82825	.2074	.2985	.5677	22
39	.63810	.76996	.82874	.2066	.2988	.5672	21
40	.63832	.76977	.82923	1.2059	1.2991	1.5666	20
41	.63854	.76958	.82972	.2052	.2994	.5661	19
42	.63877	.76940	.83022	.2045	.2997	.5655	18
43	.63899	.76921	.83071	.2038	.3000	.5650	17
44	.63921	.76903	.83120	.2031	.3003	.5644	16
45	.63944	.76884	.83169	1.2024	1.3006	1.5639	15
46	.63966	.76865	.83218	.2016	.3010	.5633	14
47	.63989	.76847	.83267	.2009	.3013	.5628	13
48	.64011	.76828	.83317	.2002	.3016	.5622	12
49	.64033	.76810	.83366	.1995	.3019	.5617	11
50	.64056	.76791	.83415	1.1988	1.3027	1.5611	10
51	.64078	.76772	.83465	.1981	.3025	.5606	9
52	.64100	.76754	.83514	.1974	.3029	.5600	8
53	.64123	.76735	.83563	.1967	.3032	.5595	7
54	.64145	.76716	.83613	.1960	.3035	.5590	6
55	.64167	.76698	.83662	1.1953	1.3038	1.5584	5
56	.64189	.76679	.83712	.1946	.3041	.5579	4
57	.64212	.76660	.83761	.1939	.3044	.5573	3
58	.64234	.76642	.83811	.1932	.3048	.5568	2
59	.64256	.76623	.83860	.1924	.3051	.5563	1
60	.64279	.76604	.83910	1.1917	1.3054	1.5557	0
M	Cosine	Sine	Cotan.	Tan.	Cosec.	Secant	M

M	Sine	Cosine	Tan.	Cotan.	Secant	Cosec.	M
0	.64279	.76604	.83910	1.1917	1.3054	1.5557	60
1	.64301	.76586	.83959	.1910	.3057	.5552	59
2	.64323	.76567	.84009	.1903	.3060	.5546	58
3	.64345	.76548	.84059	.1896	.3064	.5541	57
4	.64368	.76530	.84108	.1889	.3067	.5536	56
5	.64390	.76511	.84158	1.1882	1.3070	1.5530	55
6	.64412	.76492	.84208	.1875	.3073	.5525	54
7	.64435	.76473	.84257	.1868	.3076	.5520	53
8	.64457	.76455	.84307	.1861	.3080	.55[illegible]	52
9	.64479	.76436	.84357	.1854	.3083	.5509	51
10	.64501	.76417	.84407	1.1847	1.3086	1.5503	50
11	.64523	.76398	.84457	.1840	.3089	.5498	49
12	.64546	.76380	.84506	.1833	.3092	.5493	48
13	.64568	.76361	.84556	.1826	.3096	.5487	47
14	.64590	.76342	.84606	.1819	.3099	.5482	46
15	.64612	.76323	.84656	1.1812	1.3102	1.5477	45
16	.64635	.76304	.84706	.1805	.3105	.5471	44
17	.64657	.76286	.84756	.1798	.3109	.5466	43
18	.64679	.76267	.84806	.1791	.3112	.5461	42
19	.64701	.76248	.84856	.1785	.3115	.5456	41
20	.64723	.76229	.84906	1.1778	1.3118	1.5450	40
21	.64745	.76210	.84956	.1771	.3121	.5445	39
22	.64768	.76191	.85006	.1764	.3125	.5440	38
23	.64790	.76173	.85056	.1757	.3128	.5434	37
24	.64812	.76154	.85107	.1750	.3131	.5429	36
25	.64834	.76135	.85157	1.1743	1.3134	1.5424	35
26	.64856	.76116	.85207	.1736	.3138	.5419	34
27	.64878	.76097	.85257	.1729	.3141	.5413	33
28	.64900	.76078	.85307	.1722	.3144	.5408	32
29	.64923	.76059	.85358	.1715	.3148	.5403	31
30	.64945	.76041	.85408	1.1708	1.3151	1.5398	30
31	.64967	.76022	.85458	.1702	.3154	.5392	29
32	.64989	.76003	.85509	.1695	.3157	.5387	28
33	.65011	.75984	.85559	.1688	.3161	.5382	27
34	.65033	.75965	.85609	.1681	.3164	.5377	26
35	.65055	.75946	.85660	1.1674	1.3167	1.5371	25
36	.65077	.75927	.85710	.1667	.3170	.5366	24
37	.65100	.75908	.85761	.1660	.3174	.5361	23
38	.65121	.75889	.85811	.1653	.3177	.5356	22
39	.65144	.75870	.85862	.1647	.3180	.5351	21
40	.65166	.75851	.85912	1.1640	1.3184	1.5345	20
41	.65188	.75832	.85963	.1633	.3187	.5340	19
42	.65210	.75813	.86013	.1626	.3190	.5335	18
43	.65232	.75794	.86064	.1619	.3193	.5330	17
44	.65254	.75775	.86115	.1612	.3197	.5325	16
45	.65276	.75756	.86165	1.1605	1.3200	1.5319	15
46	.65298	.75737	.86216	.1599	.3203	.5314	14
47	.65320	.75718	.86267	.1592	.3207	.5309	13
48	.65342	.75700	.86318	.1585	.3210	.5304	12
49	.65364	.75680	.86368	.1578	.3213	.5290	11
50	.65386	.75661	.86419	1.1571	1.3217	1.5294	10
51	.65408	.75642	.86470	.1565	.3220	.5289	9
52	.65430	.75623	.86521	.1558	.3223	.5283	8
53	.65452	.75604	.86572	.1551	.3227	.5278	7
54	.65474	.75585	.86623	.1544	.3230	.5273	6
55	.65496	.75566	.86674	1.1537	1.3233	1.5268	5
56	.65518	.75547	.86725	.1531	.3237	.5263	4
57	.65540	.75528	.86775	.1524	.3240	.5258	3
58	.65562	.75509	.86826	.1517	.3243	.5253	2
59	.65584	.75490	.86878	.1510	.3247	.5248	1
60	.65606	.75471	.86929	1.1504	1.3250	1.5242	0
M	Cosine	Sine	Cotan.	Tan.	Cosec.	Secant	M

M	Sine	Cosine	Tan.	Cotan.	Secant	Cosec.	M
0	.65606	.75471	.86929	1.1504	1.3250	1.5242	60
1	.65628	.75452	.86980	.1497	.3253	.5237	59
2	.65650	.75433	.87031	.1490	.3257	.5232	58
3	.65672	.75414	.87082	.1483	.3260	.5227	57
4	.65694	.75394	.87133	.1477	.3263	.5222	56
5	.65716	.75375	.87184	1.1470	1.3267	1.5217	55
6	.65737	.75356	.87235	.1463	.3270	.5212	54
7	.65759	.75337	.87287	.1456	.3274	.5207	53
8	.65781	.75318	.87338	.1450	.3277	.5202	52
9	.65803	.75299	.87389	.1443	.3280	.5197	51
10	.65825	.75280	.87441	1.1436	1.3284	1.5192	50
11	.65847	.75261	.87492	.1430	.3287	.5187	49
12	.65869	.75241	.87543	.1423	.3290	.5182	48
13	.65891	.75222	.87595	.1416	.3294	.5177	47
14	.65913	.75203	.87646	.1409	.3297	.5171	46
15	.65934	.75184	.87698	1.1403	1.3301	1.5166	45
16	.65956	.75165	.87749	.1396	.3304	.5161	44
17	.65978	.75146	.87801	.1389	.3307	.5156	43
18	.66000	.75126	.87852	.1383	.3311	.5151	42
19	.66022	.75107	.87904	.1376	.3314	.5146	41
20	.66044	.75088	.87955	1.1369	1.3318	1.5141	40
21	.66066	.75069	.88007	.1363	.3321	.5136	39
22	.66087	.75049	.88058	.1356	.3324	.5131	38
23	.66109	.75030	.88110	.1349	.3328	.5126	37
24	.66131	.75011	.88162	.1343	.3331	.5121	36
25	.66153	.74992	.88213	1.1336	1.3335	1.5116	35
26	.66175	.74973	.88265	.1329	.3338	.5111	34
27	.66197	.74953	.88317	.1323	.3342	.5106	33
28	.66218	.74934	.88369	.1316	.3345	.5101	32
29	.66240	.74915	.88421	.1309	.3348	.5096	31
30	.66262	.74895	.88472	1.1303	1.3352	1.5092	30
31	.66284	.74876	.88524	.1296	.3355	.5087	29
32	.66305	.74857	.88576	.1290	.3359	.5082	28
33	.66327	.74838	.88628	.1283	.3362	.5077	27
34	.66349	.74818	.88680	.1276	.3366	.5072	26
35	.66371	.74799	.88732	1.1270	1.3369	1.5067	25
36	.66393	.74780	.88784	.1263	.3372	.5062	24
37	.66414	.74760	.88836	.1257	.3376	.5057	23
38	.66436	.74741	.88888	.1250	.3379	.5052	22
39	.66458	.74722	.88940	.1243	.3383	.5047	21
40	.66479	.74702	.88992	1.1237	1.3386	1.5042	20
41	.66501	.74683	.89044	.1230	.3390	.5037	19
42	.66523	.74664	.89097	.1224	.3393	.5032	18
43	.66545	.74644	.89149	.1217	.3397	.5027	17
44	.66566	.74625	.89201	.1211	.3400	.5022	16
45	.66588	.74606	.89253	1.1204	1.3404	1.5018	15
46	.66610	.74586	.89306	.1197	.3407	.5013	14
47	.66631	.74567	.89358	.1191	.3411	.5008	13
48	.66653	.74548	.89410	.1184	.3414	.5003	12
49	.66675	.74528	.89463	.1178	.3418	.4998	11
50	.66697	.74509	.89515	1.1171	1.3421	1.4993	10
51	.66718	.74489	.89567	.1165	.3425	.4988	9
52	.66740	.74470	.89620	.1158	.3428	.4983	8
53	.66762	.74450	.89672	.1152	.3432	.4979	7
54	.66783	.74431	.89725	.1145	.3435	.4974	6
55	.66805	.74412	.89777	1.1139	1.3439	1.4969	5
56	.66826	.74392	.89830	.1132	.3442	.4964	4
57	.66848	.74373	.89882	.1126	.3446	.4959	3
58	.66870	.74353	.89935	.1119	.3449	.4954	2
59	.66891	.74334	.89988	.1113	.3453	.4949	1
60	.66913	.74314	.90040	1.1106	1.3456	1.4945	0
M	Cosine	Sine	Cotan.	Tan.	Cosec.	Secant	M

M	Sine	Cosine	Tan.	Cotan.	Secant	Cosec.	M
0	.66913	.74314	.90040	1.1106	1.3456	1.4945	60
1	.66935	.74295	.90093	.1100	.3460	.4940	59
2	.66956	.74275	.90146	.1093	.3463	.4935	58
3	.66978	.74256	.90198	.1086	.3467	.4930	57
4	.66999	.74236	.90251	.1080	.3470	.4925	56
5	.67021	.74217	.90304	1.1074	1.3474	1.4921	55
6	.67043	.74197	.90357	.1067	.3477	.4916	54
7	.67064	.74178	.90410	.1061	.3481	.4911	53
8	.67086	.74158	.90463	.1054	.3485	.4906	52
9	.67107	.74139	.90515	.1048	.3488	.4901	51
10	.67129	.74119	.90568	1.1041	1.3492	1.4897	50
11	.67150	.74100	.90621	.1035	.3495	.4892	49
12	.67172	.74080	.90674	.1028	.3499	.4887	48
13	.67194	.74061	.90727	.1022	.3502	.4882	47
14	.67215	.74041	.90780	.1015	.3506	.4877	46
15	.67237	.74022	.90834	1.1009	1.3509	1.4873	45
16	.67258	.74002	.90887	.1003	.3513	.4868	44
17	.67280	.73983	.90940	.0996	.3517	.4863	43
18	.67301	.73963	.90993	.0990	.3520	.4858	42
19	.67323	.73943	.91046	.0983	.3524	.4854	41
20	.67344	.73924	.91099	1.0977	1.3527	1.4849	40
21	.67366	.73904	.91153	.0971	.3531	.4844	39
22	.67387	.73885	.91206	.0964	.3534	.4839	38
23	.67409	.73865	.91259	.0958	.3538	.4835	37
24	.67430	.73845	.91312	.0951	.3542	.4830	36
25	.67452	.73826	.91366	1.0945	1.3545	1.4825	35
26	.67473	.73806	.91419	.0939	.3549	.4821	34
27	.67495	.73787	.91473	.0932	.3552	.4816	33
28	.67516	.73767	.91526	.0926	.3556	.4811	32
29	.67537	.73747	.91580	.0919	.3560	.4806	31
30	.67559	.73728	.91633	1.0913	1.3563	1.4802	30
31	.67580	.73708	.91687	.0907	.3567	.4797	29
32	.67602	.73688	.91740	.0900	.3571	.4792	28
33	.67623	.73669	.91794	.0894	.3574	.4788	27
34	.67645	.73649	.91847	.0888	.3578	.4783	26
35	.67666	.73629	.91901	1.0881	1.3581	1.4778	25
36	.67688	.73610	.91955	.0875	.3585	.4774	24
37	.67709	.73590	.92008	.0868	.3589	.4769	23
38	.67730	.73570	.92062	.0862	.3592	.4764	22
39	.67752	.73551	.92116	.0856	.3596	.4760	21
40	.67773	.73531	.92170	1.0849	1.3600	1.4755	20
41	.67794	.73511	.92223	.0843	.3603	.4750	19
42	.67816	.73491	.92277	.0837	.3607	.4746	18
43	.67837	.73472	.92331	.0830	.3611	.4741	17
44	.67859	.73452	.92385	.0824	.3614	.4736	16
45	.67880	.73432	.92439	1.0818	1.3618	1.4732	15
46	.67901	.73412	.92493	.0812	.3622	.4727	14
47	.67923	.73393	.92547	.0805	.3625	.4723	13
48	.67944	.73373	.92601	.0799	.3629	.4718	12
49	.67965	.73353	.97655	.0793	.3633	.4713	11
50	.67987	.73333	.92709	1.0786	1.3636	1.4709	10
51	.68008	.73314	.92763	.0780	.3640	.4704	9
52	.68029	.73294	.92817	.0774	.3644	.4699	8
53	.68051	.73274	.92871	.0767	.3647	.4695	7
54	.68072	.73254	.92926	.0761	.3651	.4690	6
55	.68093	.73234	.92980	1.0755	1.3655	1.4686	5
56	.68115	.73215	.93034	.0749	.3658	.4681	4
57	.68136	.73195	.93088	.0742	.3662	.4676	3
58	.68157	.73175	.93143	.0736	.3666	.4672	2
59	.68178	.73155	.93197	.0730	.3669	.4667	1
60	.68200	.73135	.93251	1.0724	1.3673	1.4663	0
M	Cosine	Sine	Cotan.	Tan.	Cosec.	Secant	M

43°

M	Sine	Cosine	Tan.	Cotan.	Secant	Cosec.	M
0	.68200	.73135	.93251	1.0724	1.3673	1.4663	60
1	.68221	.73115	.93306	.0717	.3677	.4658	59
2	.68242	.73096	.93360	.0711	.3681	.4654	58
3	.68264	.73076	.93415	.0705	.3684	.4649	57
4	.68285	.73056	.93469	.0699	.3688	.4644	56
5	.68306	.73036	.93524	1.0692	1.3692	1.4640	55
6	.68327	.73016	.93578	.0686	.3695	.4635	54
7	.68349	.72996	.93633	.0680	.3699	.4631	53
8	.68370	.72976	.93687	.0674	.3703	.4626	52
9	.68391	.72956	.93742	.0667	.3707	.4622	51
10	.68412	.72937	.93797	1.0661	1.3710	1.4617	50
11	.68433	.72917	.93851	.0655	.3714	.4613	49
12	.68455	.72897	.93906	.0649	.3718	.4608	48
13	.68476	.72877	.93961	.0643	.3722	.4604	47
14	.68497	.72857	.94016	.0636	.3725	.4599	46
15	.68518	.72837	.94071	1.0630	1.3729	1.4595	45
16	.68539	.72817	.94125	.0624	.3733	.4590	44
17	.68561	.72797	.94180	.0618	.3737	.4586	43
18	.68582	.72777	.94235	.0612	.3740	.4581	42
19	.68603	.72757	.94290	.0605	.3744	.4577	41
20	.68624	.72737	.94345	1.0599	1.3748	1.4572	40
21	.68645	.72717	.94400	.0593	.3752	.4568	39
22	.68666	.72697	.94455	.0587	.3756	.4563	38
23	.68688	.72677	.94510	.0581	.3759	.4559	37
24	.68709	.72657	.94565	.0575	.3763	.4554	36
25	.68730	.72637	.94620	1.0568	1.3767	1.4550	35
26	.68751	.72617	.94675	.0562	.3771	.4545	34
27	.68772	.72597	.94731	.0556	.3774	.4541	33
28	.68793	.72577	.94786	.0550	.3778	.4536	32
29	.68814	.72557	.94841	.0544	.3782	.4532	31
30	.68835	.72537	.94896	1.0538	1.3786	1.4527	30
31	.68856	.72517	.94952	.0532	.3790	.4523	29
32	.68878	.72497	.95007	.0525	.3794	.4518	28
33	.68899	.72477	.95062	.0519	.3797	.4514	27
34	.68920	.72457	.95118	.0513	.3801	.4510	26
35	.68941	.72437	.95173	1.0507	1.3805	1.4505	25
36	.68962	.72417	.95229	.0501	.3809	.4501	24
37	.68983	.72397	.95284	.0495	.3813	.4496	23
38	.69004	.72377	.95340	.0489	.3816	.4492	22
39	.69025	.72357	.95395	.0483	.3820	.4487	21
40	.69046	.72337	.95451	1.0476	1.3824	1.4483	20
41	.69067	.72317	.95506	.0470	.3828	.4479	19
42	.69088	.72297	.95562	.0464	.3832	.4474	18
43	.69109	.72277	.95618	.0458	.3836	.4470	17
44	.69130	.72256	.95673	.0452	.3839	.4465	16
45	.69151	.72236	.95729	1.0446	1.3843	1.4461	15
46	.69172	.72216	.95785	.0440	.3847	.4457	14
47	.69193	.72196	.95841	.0434	.3851	.4452	13
48	.69214	.72176	.95896	.0428	.3855	.4448	12
49	.69235	.72156	.95952	.0422	.3859	.4443	11
50	.69256	.72136	.96008	1.0416	1.3863	1.4439	10
51	.69277	.72115	.96064	.0410	.3867	.4435	9
52	.69298	.72095	.96120	.0404	.3870	.4430	8
53	.69319	.72075	.96176	.0397	.3874	.4426	7
54	.69340	.72055	.96232	.0391	.3878	.4422	6
55	.69361	.72035	.96288	1.0385	1.3882	1.4417	5
56	.69382	.72015	.96344	.0379	.3886	.4413	4
57	.69403	.71994	.96400	.0373	.3890	.4408	3
58	.69424	.71974	.96456	.0367	.3894	.4404	2
59	.69445	.71954	.96513	.0361	.3898	.4400	1
60	.69466	.71934	.96569	1.0355	1.3902	1.4395	0
M	Cosine	Sine	Cotan.	Tan.	Cosec.	Secant	M

46°

44°

M	Sine	Cosine	Tan.	Cotan.	Secant	Cosec.	M
0	.69466	.71934	.96569	1.0355	1.3902	1.4395	60
1	.69487	.71914	.96625	.0349	.3905	.4391	59
2	.69508	.71893	.96681	.0343	.3909	.4387	58
3	.69528	.71873	.96738	.0337	.3913	.4382	57
4	.69549	.71853	.96794	.0331	.3917	.4378	56
5	.69570	.71833	.96850	1.0325	1.3921	1.4374	55
6	.69591	.71813	.96907	.0319	.3925	.4370	54
7	.69612	.71792	.96963	.0313	.3929	.4365	53
8	.69633	.71772	.97020	.0307	.3933	.4361	52
9	.69654	.71752	.97076	.0301	.3937	.4357	51
10	.69675	.71732	.97133	1.0295	1.3941	1.4352	50
11	.69696	.71711	.97189	.0289	.3945	.4348	49
12	.69716	.71691	.97246	.0283	.3949	.4344	48
13	.69737	.71671	.97302	.0277	.3953	.4339	47
14	.69758	.71650	.97359	.0271	.3957	.4335	46
15	.69779	.71630	.97416	1.0265	1.3960	1.4331	45
16	.69800	.71610	.97472	.0259	.3964	.4327	44
17	.69821	.71589	.97529	.0253	.3968	.4322	43
18	.69841	.71569	.97586	.0247	.3972	.4318	42
19	.69862	.71549	.97643	.0241	.3976	.4314	41
20	.69883	.71529	.97700	1.0235	1.3980	1.4310	40
21	.69904	.71508	.97756	.0229	.3984	.4305	39
22	.69925	.71488	.97813	.0223	.3988	.4301	38
23	.69945	.71468	.97870	.0218	.3992	.4297	37
24	.69966	.71447	.97927	.0212	.3996	.4292	36
25	.69987	.71427	.97984	1.0206	1.4000	1.4288	35
26	.70008	.71406	.98041	.0200	.4004	.4284	34
27	.70029	.71386	.98098	.0194	.4008	.4280	33
28	.70049	.71366	.98155	.0188	.4012	.4276	32
29	.70070	.71345	.98212	.0182	.4016	.4271	31
30	.70091	.71325	.98270	1.0176	1.4020	1.4267	30
31	.70112	.71305	.98327	.0170	.4024	.4263	29
32	.70132	.71284	.98384	.0164	.4028	.4259	28
33	.70153	.71264	.98441	.0158	.4032	.4254	27
34	.70174	.71243	.98499	.0152	.4036	.4250	26
35	.70194	.71223	.98556	1.0146	1.4040	1.4246	25
36	.70215	.71203	.98613	.0141	.4044	.4242	24
37	.70236	.71182	.98671	.0135	.4048	.4238	23
38	.70257	.71162	.98728	.0129	.4052	.4233	22
39	.70277	.71141	.98786	.0123	.4056	.4229	21
40	.70298	.71121	.98843	1.0117	1.4060	1.4225	20
41	.70319	.71100	.98901	.0111	.4065	.4221	19
42	.70339	.71080	.98958	.0105	.4069	.4217	18
43	.70360	.71059	.99016	.0099	.4073	.4212	17
44	.70381	.71039	.99073	.0093	.4077	.4208	16
45	.70401	.71018	.99131	1.0088	1.4081	1.4204	15
46	.70422	.70998	.99189	.0082	.4085	.4200	14
47	.70443	.70977	.99246	.0076	.4089	.4196	13
48	.70463	.70957	.99304	.0070	.4093	.4192	12
49	.70484	.70936	.99362	.0064	.4097	.4188	11
50	.70505	.70916	.99420	1.0058	1.4101	1.4183	10
51	.70525	.70895	.99478	.0052	.4105	.4179	9
52	.70546	.70875	.99536	.0047	.4109	.4175	8
53	.70566	.70854	.99593	.0041	.4113	.4171	7
54	.70587	.70834	.99651	.0035	.4117	.4167	6
55	.70608	.70813	.99709	1.0029	1.4122	1.4163	5
56	.70628	.70793	.99767	.0023	.4126	.4159	4
57	.70649	.70772	.99826	.0017	.4130	.4154	3
58	.70669	.70752	.99884	.0012	.4134	.4150	2
59	.70690	.70731	.99942	.0006	.4138	.4146	1
60	.70711	.70711	1.0000	1.0000	1.4142	1.4142	0
M	Cosine	Sine	Cotan.	Tan.	Cosec.	Secant	M

45°

127

DECIMAL EQUIVALENTS

OF COMMON FRACTIONS OF AN INCH

		$\frac{1}{64}$	.01563			$\frac{33}{64}$	.51563
	$\frac{1}{32}$		.03125		$\frac{17}{32}$		.53125
		$\frac{3}{64}$	.04688			$\frac{35}{64}$	.54688
$\frac{1}{16}$			.0625	$\frac{9}{16}$			.5625
		$\frac{5}{64}$	.07813			$\frac{37}{64}$	.57813
	$\frac{3}{32}$		.09375		$\frac{19}{32}$		.59375
		$\frac{7}{64}$	.10938			$\frac{39}{64}$	.60938
$\frac{1}{8}$			.125	$\frac{5}{8}$			.625
		$\frac{9}{64}$	.14063			$\frac{41}{64}$	.64063
	$\frac{5}{32}$		.15625		$\frac{21}{32}$		.65625
		$\frac{11}{64}$	.17188			$\frac{43}{64}$	.67188
$\frac{3}{16}$			.1875	$\frac{11}{16}$			.6875
		$\frac{13}{64}$	.20313			$\frac{45}{64}$	.70313
	$\frac{7}{32}$		.21875		$\frac{23}{32}$		.71875
		$\frac{15}{64}$	.23438			$\frac{47}{64}$	.73438
$\frac{1}{4}$			.250	$\frac{3}{4}$			.750
		$\frac{17}{64}$	.26563			$\frac{49}{64}$	.76563
	$\frac{9}{32}$		.28125		$\frac{25}{32}$		.78125
		$\frac{19}{64}$	.29688			$\frac{51}{64}$	.79688
$\frac{5}{16}$			.3125	$\frac{13}{16}$			.8125
		$\frac{21}{64}$	.32813			$\frac{53}{64}$	.82813
	$\frac{11}{32}$		.34375		$\frac{27}{32}$		.84375
		$\frac{23}{64}$	.35938			$\frac{55}{64}$	.85938
$\frac{3}{8}$			.375	$\frac{7}{8}$			.875
		$\frac{25}{64}$	.39063			$\frac{57}{64}$	.89063
	$\frac{13}{32}$		.40625		$\frac{29}{32}$		.90625
		$\frac{27}{64}$	.42188			$\frac{59}{64}$	.92188
$\frac{7}{16}$			.4375	$\frac{15}{16}$			.9375
		$\frac{29}{64}$	.45313			$\frac{61}{64}$	.95313
	$\frac{15}{32}$		.46875		$\frac{31}{32}$		.96875
		$\frac{31}{64}$	.48438			$\frac{63}{64}$	.98438
$\frac{1}{2}$			.500	1			1.00000

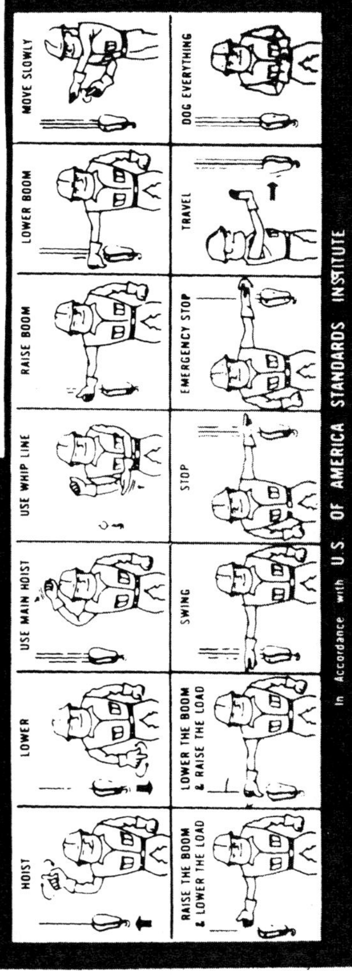

CRANE SIGNALS
HOIST
LOWER
USE MAIN HOIST
USE WHIP LINE
RAISE BOOM
LOWER BOOM
MOVE SLOWLY
RAISE THE BOOM & LOWER THE LOAD
LOWER THE BOOM & RAISE THE LOAD
SWING
STOP
EMERGENCY STOP
TRAVEL
DOG EVERYTHING
In Accordance with U.S. OF AMERICA STANDARDS INSTITUTE

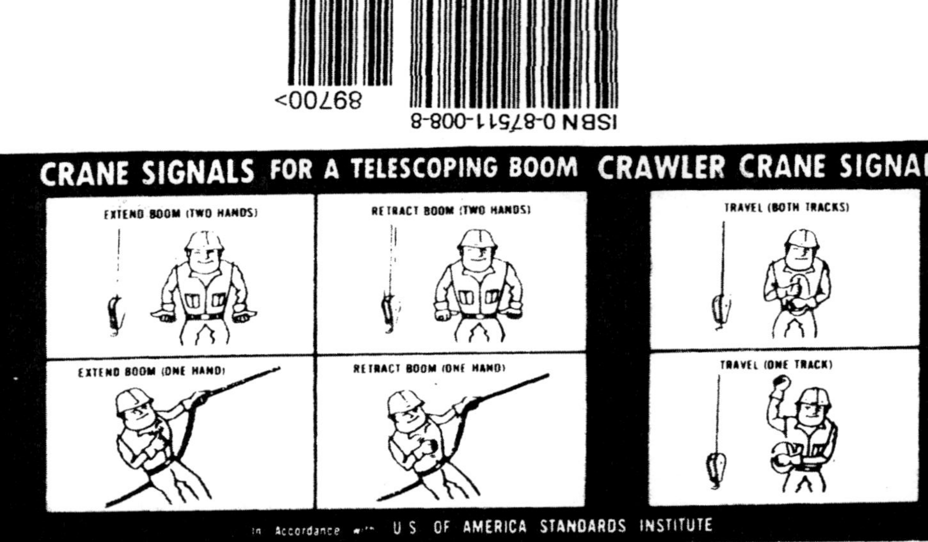

ISBN 0-87511-008-8
9 780875 110080
89700>